Houghton
Mifflin
Harcourt

D0127979

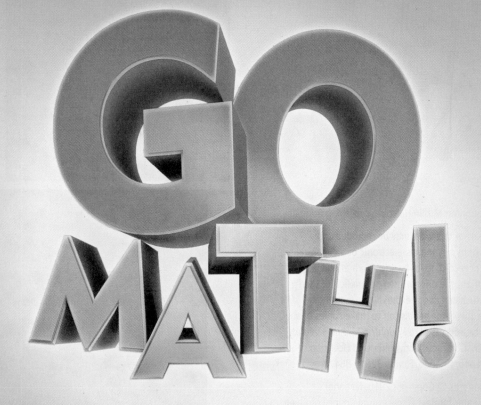

Made in the United States
Text printed on 100%
recycled paper

Houghton
Mifflin
Harcourt

Printed in the U.S.A.

ISBN 978-0-544-34230-9

17   0928   22 21 20 19

4500746793   C D E F G

Dear Students and Families,

Welcome to **Go Math!**, Grade 4! In this exciting mathematics program, there are hands-on activities to do and real-world problems to solve. Best of all, you will write your ideas and answers right in your book. In **Go Math!**, writing and drawing on the pages helps you think deeply about what you are learning, and you will really understand math!

By the way, all of the pages in your **Go Math!** book are made using recycled paper. We wanted you to know that you can Go Green with **Go Math!**

Sincerely,

The Authors

Made in the United States
Text printed on 100% recycled paper

GO MATH!

Authors

Juli K. Dixon, Ph.D.
Professor, Mathematics Education
University of Central Florida
Orlando, Florida

Edward B. Burger, Ph.D.
President, Southwestern University
Georgetown, Texas

Steven J. Leinwand
Principal Research Analyst
American Institutes for
  Research (AIR)
Washington, D.C.

Matthew R. Larson, Ph.D.
K-12 Curriculum Specialist for
  Mathematics
Lincoln Public Schools
Lincoln, Nebraska

Martha E. Sandoval-Martinez
Math Instructor
El Camino College
Torrance, California

Contributor

Rena Petrello
Professor, Mathematics
Moorpark College
Moorpark, California

English Language Learners Consultant

Elizabeth Jiménez
CEO, GEMAS Consulting
Professional Expert on English
  Learner Education
Bilingual Education and
  Dual Language
Pomona, California

Geometry, Measurement, and Data

Critical Area Understanding that geometric figures can be analyzed and classified based on their properties, such as having parallel sides, perpendicular sides, particular angle measures, and symmetry

Critical Area

GO DIGITAL

Go online! Your math lessons are interactive. Use *i*Tools, Animated Math Models, the Multimedia eGlossary, and more.

Chapter 12 Overview

In this chapter, you will explore and discover answers to the following **Essential Questions**:

• How can you use relative sizes of measurements to solve problems and to generate measurement tables that show a relationship?

• How can you compare metric units of length, mass, or liquid volume?

• How can you compare customary units of length, weight, or liquid volume?

Personal Math Trainer
Online Assessment and Intervention

FOR MORE PRACTICE
GO TO THE
Personal Math Trainer

Practice and Homework

Lesson Check and
Spiral Review in
every lesson

Relative Sizes of Measurement Units

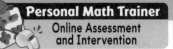

✓ Show What You Know

Personal Math Trainer
Online Assessment and Intervention

Check your understanding of important skills.

Name _____

▶ **Time to the Half Hour** **Read the clock. Write the time.** (2.MD.C.7)

1. _____   2. _____   3. _____

▶ **Multiply by 1-Digit Numbers** **Find the product.** (4.NBT.B.5)

4. 84
   × 7

5. 536
   × 8

6. 748
   × 5

7. 2,524
   × 2

8. 360
   × 9

9. 296
   × 3

10. $1,428
   × 4

11. 64
   × 5

Math in the Real World

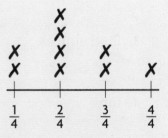

A team was given a bucket of water and a sponge. The team had 1 minute to fill an empty half-gallon bucket with water using only the sponge. The line plot shows the amount of water squeezed into the bucket. Did the team squeeze enough water to fill the half-gallon bucket?

Amount of Water Squeezed into the Bucket (in cups)

Vocabulary Builder

▶ **Visualize It** ••••••••••••••••••••••••••••••••

Complete the Brain Storming diagram by using words with a ✓.

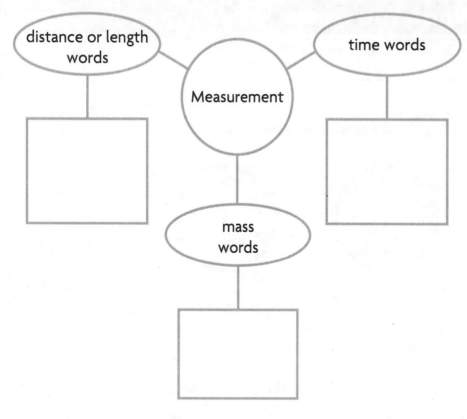

▶ **Understand Vocabulary** ••••••••••••••••••••••••••

Draw a line to match each word with its definition.

1. decimeter

2. second

3. fluid ounce

4. ton

5. line plot

- A customary unit for measuring liquid volume

- A graph that shows the frequency of data along a number line

- A customary unit used to measure weight

- A small unit of time

- A metric unit for measuring length or distance

 GO DIGITAL
- Interactive Student Edition
- Multimedia eGlossary

Chapter 12 Vocabulary

cup (c)

taza (tz)

18

fluid ounce (fl oz)

onza fluida (fl oz)

34

gallon (gal)

galón (gal)

37

half gallon

medio galón

38

kilometer (km)

kilómetro (km)

44

line plot

diagrama de puntos

47

liquid volume

volumen de un líquido

49

mile (mi)

milla (mi)

51

A customary unit used to measure liquid capacity and liquid volume

CUP

1 cup = 8 fluid ounces

A customary unit used to measure capacity and liquid volume
1 cup = 8 ounces

CUP

A customary unit for measuring capacity and liquid volume
1 half gallon = 2 quarts

Milk 1 half gallon

A customary unit for measuring capacity and liquid volume
1 gallon = 4 quarts

1 gallon

A graph that records each piece of data on a number line

Example:

Height of Bean Seedlings

A metric unit for measuring length or distance
1 kilometer = 1,000 meters

A customary unit for measuring length or distance
1 mile = 5,280 feet

The measure of the space a liquid occupies

1 cup = 8 fluid ounces

1 pint = 2 cups

1 quart = 4 cups

Metric Units of Liquid Volume
1 liter (L) = 1,000 milliliters (mL)

milliliter (mL)

mililitro (mL)

52

millimeter (mm)

milímetro (mm)

53

ounce (oz)

onza (oz)

58

pint (pt)

pinta (pt)

67

pound (lb)

libra (lb)

70

quart (qt)

cuarto (ct)

74

second (sec)

segundo (seg)

83

ton (T)

tonelada (t)

92

A metric unit for measuring length or distance
1 centimeter = 10 millimeters

centimeters

A metric unit for measuring capacity and liquid volume
1 liter = 1,000 milliliters

1 milliliter

A customary unit for measuring capacity and liquid volume
1 pint = 2 cups

1 pint

A customary unit for measuring weight
1 pound = 16 ounces

about 1 ounce

A customary unit for measuring capacity and liquid volume
1 quart = 2 pints

A customary unit for measuring weight
1 pound = 16 ounces

about 1 pound

A customary unit used to measure weight
1 ton = 2,000 pounds

about 1 ton

A small unit of time
1 minute = 60 seconds

1 second

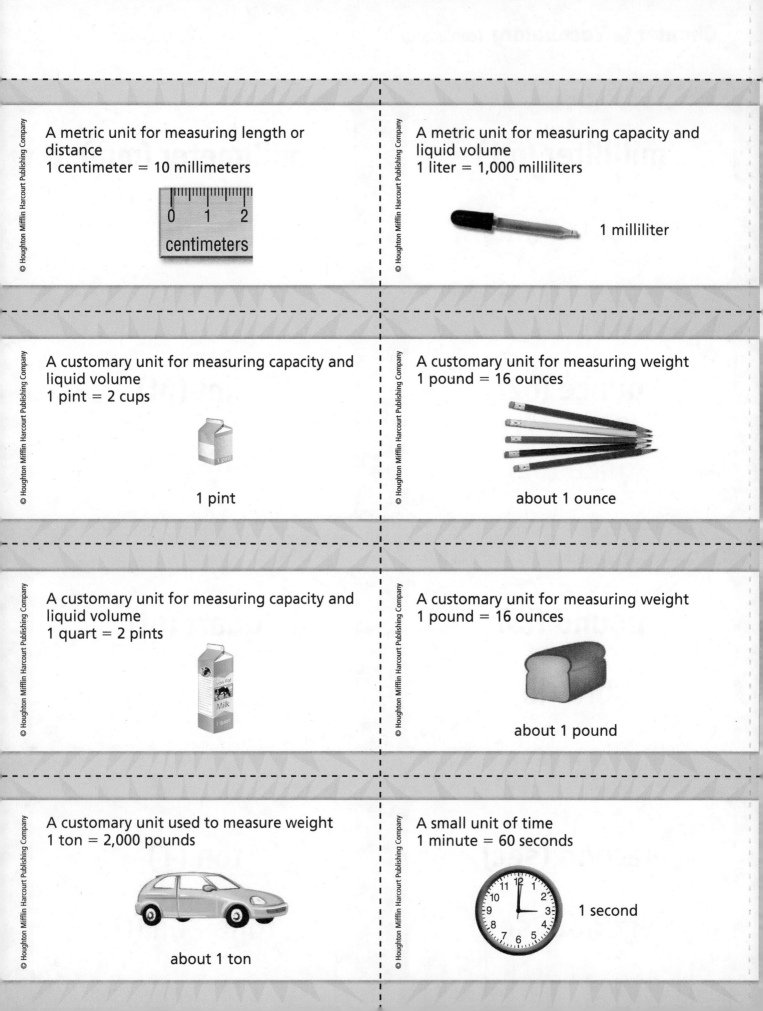

Game

Bingo

For 3 to 6 players

Materials

- 1 set of word cards
- 1 Bingo board for each player
- game markers

How to Play

1. The caller chooses a card and reads the definition. Then the caller puts the card in a second pile.
2. Players put a marker on the word that matches the definition each time they find it on their Bingo boards.
3. Repeat Steps 1 and 2 until a player marks 5 boxes in a line going down, across, or on a slant and calls "Bingo."
4. Check the answers. Have the player who said "Bingo" read the words aloud while the caller checks the definitions on the cards in the second pile.

Word Box
cup
fluid ounce
gallon
half gallon
kilometer
line plot
liquid volume
mile
milliliter
millimeter
ounce
pint
pound
quart
second
ton

The Write Way

Reflect

Choose one idea. Write about it.

- Do 50 milliliters and 50 millimeters represent the same amount? Explain why or why not.
- Write a paragraph that uses at least three of these words.

  cup   mile   pound   second   ton

- Explain what is most important to understand about line plots.

Measurement Benchmarks

Essential Question How can you use benchmarks to understand the relative sizes of measurement units?

Common Core Measurement and Data—
4.MD.A.1
MATHEMATICAL PRACTICES
MP1, MP2, MP3, MP7

🔑 Unlock the Problem · Real World

Jake says the length of his bike is about four yards. Use the benchmark units below to determine if Jake's statement is reasonable.

Customary Units of Length

1 in. about 1 inch	└ 1 ft ┘ about 1 foot	└ 1 yd ┘ about 1 yard	1 mile in about 20 minutes

A **mile** is a customary unit for measuring length or distance. The benchmark shows the distance you can walk in about 20 minutes.

A baseball bat is about one yard long. Since Jake's bike is shorter than four times the length of a baseball bat, his bike is shorter than four yards long.

So, Jake's statement _____ reasonable.

Jake's bike is about _____ baseball bats long.

🔑 Example 1 Use the benchmark customary units.

Customary Units of Liquid Volume

CUP			Milk	
1 cup = 8 fluid ounces	1 pint	1 quart	1 half gallon	1 gallon

• About how much liquid is in a mug of hot chocolate? _____

Customary Units of Weight

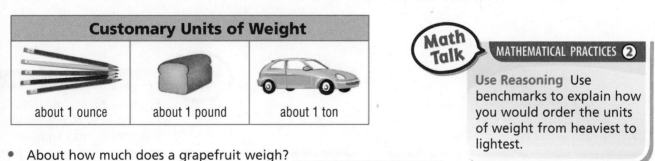

about 1 ounce	about 1 pound	about 1 ton

Math Talk MATHEMATICAL PRACTICES ②

Use Reasoning Use benchmarks to explain how you would order the units of weight from heaviest to lightest.

• About how much does a grapefruit weigh? _____

Benchmarks for Metric Units Like place value, the metric system
is based on multiples of ten. Each unit is 10 times as large as the next
smaller unit. Below are some common metric benchmarks.

🔑 **Example 2** Use the benchmark metric units.

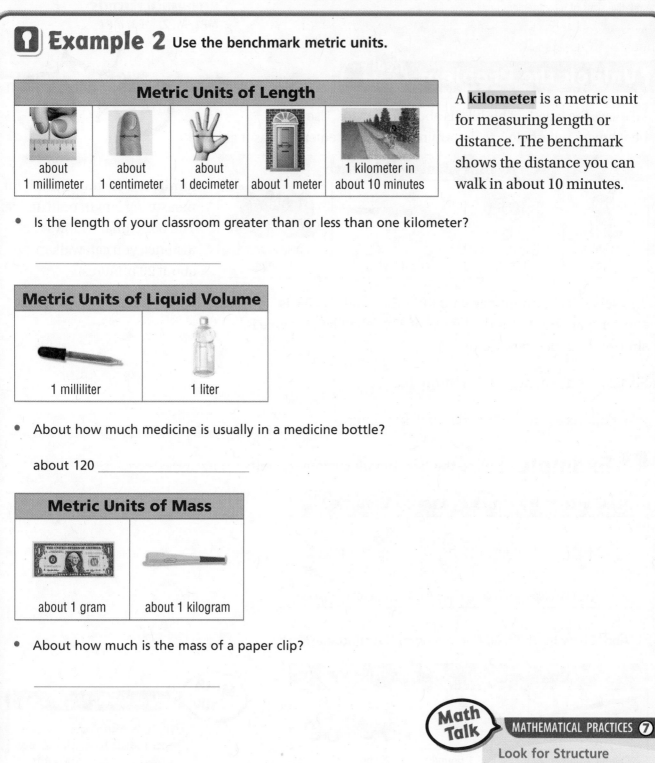

Metric Units of Length				
about 1 millimeter	about 1 centimeter	about 1 decimeter	about 1 meter	1 kilometer in about 10 minutes

A **kilometer** is a metric unit for measuring length or distance. The benchmark shows the distance you can walk in about 10 minutes.

- Is the length of your classroom greater than or less than one kilometer?

_____

Metric Units of Liquid Volume	
1 milliliter	1 liter

- About how much medicine is usually in a medicine bottle?

about 120 _____

Metric Units of Mass	
about 1 gram	about 1 kilogram

- About how much is the mass of a paper clip?

_____

Math Talk

MATHEMATICAL PRACTICES ⑦

Look for Structure
Explain how benchmark measurements can help you decide which unit to use when measuring.

© Houghton Mifflin Harcourt Publishing Company

Name _____

Share and Show

Use benchmarks to choose the metric unit you would use to measure each.

1. mass of a strawberry

   _____

☑ 2. length of a cell phone

   _____

Circle the better estimate.

3. width of a teacher's desk

   10 meters or 1 meter

4. the amount of liquid a punch bowl holds

   2 liters or 20 liters

☑ 5. distance between Seattle and San Francisco

   6 miles or 680 miles

> **Math Talk** MATHEMATICAL PRACTICES ③
>
> **Apply** Which metric unit would you use to measure the distance across the United States? Explain.

Metric Units
centimeter
meter
kilometer
gram
kilogram
milliliter
liter

On Your Own

Use benchmarks to choose the customary unit you would use to measure each.

6. length of a football field

   _____

7. weight of a pumpkin

   _____

Circle the better estimate.

8. weight of a watermelon

   4 pounds or 4 ounces

9. the amount of liquid a fish tank holds

   10 cups or 10 gallons

Customary Units
inch
foot
yard
ounce
pound
cup
gallon

Complete the sentence. Write *more* or *less*.

10. Matthew's large dog weighs _____ than one ton.

11. The amount of liquid a sink can hold is _____ than one cup of water.

12. A paper clip has a mass of _____ than one kilogram.

Problem Solving • Applications (Real World)

For 13–15, use benchmarks to explain your answer.

13. **THINK SMARTER** Cristina is making macaroni and cheese for her family. Would Cristina use 1 pound of macaroni or 1 ounce of macaroni?

_____

_____

14. Which is the better estimate for the length of a kitchen table, 200 centimeters or 200 meters?

_____

_____

15. **GO DEEPER** Jodi wants to weigh her cat and measure its standing height. Which two units should she use?

_____

16. **MATHEMATICAL PRACTICE 1** **Evaluate Reasonableness** Dalton used benchmarks to estimate that there are more cups than quarts in one gallon. Is Dalton's estimate reasonable? Explain.

_____

_____

17. **THINK SMARTER** Select the correct word to complete the sentence.

Justine is thirsty after running two miles.

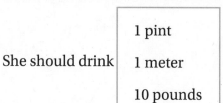

She should drink | 1 pint / 1 meter / 10 pounds | of water.

Measurement Benchmarks

COMMON CORE STANDARD—4.MD.A.1
Solve problems involving measurement and conversion of measurements from a larger unit to a smaller unit.

Use benchmarks to choose the customary unit you would use to measure each.

1. height of a computer

   foot
   _____

2. weight of a table

   _____

3. length of a semi-truck

   _____

4. the amount of liquid a bathtub holds

   _____

Customary Units	
ounce	yard
pound	mile
inch	gallon
foot	cup

Use benchmarks to choose the metric unit you would use to measure each.

5. mass of a grasshopper

   _____

6. the amount of liquid a water bottle holds

   _____

7. length of a soccer field

   _____

8. length of a pencil

   _____

Metric Units	
milliliter	centimeter
liter	meter
gram	kilometer
kilogram	

Circle the better estimate.

9. mass of a chicken egg

   50 grams   50 kilograms

10. length of a car

   12 miles   12 feet

11. amount of liquid a drinking glass holds

   8 ounces   8 quarts

Problem Solving Real World

12. What is the better estimate for the mass of a textbook, 1 gram or 1 kilogram?

   _____

13. What is the better estimate for the height of a desk, 1 meter or 1 kilometer?

   _____

14. **WRITE** *Math* Use benchmarks to determine the customary and metric units you would use to measure the height of your house. Explain your answer.

   _____

   _____

Lesson Check (4.MD.A.1)

1. What unit would be best to use for measuring the weight of a stapler?

_____

_____

2. Which is the best estimate for the length of a car?

_____

_____

Spiral Review (4.NF.B.4c, 4.NF.C.6, 4.MD.C.5a, 4.MD.C.5b, 4.G.A.2)

3. Bart practices his trumpet $1\frac{1}{4}$ hours each day. How many hours will he practice in 6 days?

_____

_____

4. Millie collected 100 stamps from different countries. Thirty-two of the stamps are from countries in Africa. What is $\frac{32}{100}$ written as a decimal?

_____

_____

5. Diedre drew a quadrilateral with 4 right angles and opposite sides of the same length. Name all the kinds of polygons that could be Diedre's quadrilateral.

_____

_____

6. How many degrees are in an angle that turns through $\frac{1}{2}$ of a circle?

_____

_____

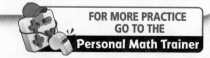

FOR MORE PRACTICE
GO TO THE
Personal Math Trainer

Name _____

Customary Units of Length

Essential Question How can you use models to compare customary units of length?

Common Core Measurement and Data—4.MD.A.1
Also 4.MD.A.2
MATHEMATICAL PRACTICES
MP2, MP3, MP4

🔑 Unlock the Problem Real World

You can use a ruler to measure length. A ruler that is 1 foot long shows 12 inches in 1 foot. A ruler that is 3 feet long is called a yardstick. There are 3 feet in 1 yard.

How does the size of a foot compare to the size of an inch?

🔓 Activity

Materials ■ 1-inch grid paper ■ scissors ■ tape

STEP 1 Cut out the paper inch tiles. Label each tile 1 inch.

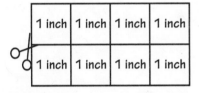

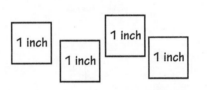

STEP 2 Place 12 tiles end-to-end to build 1 foot. Tape the tiles together.

1 foot

| 1 inch | 1 inch | 1 inch | 1 inch | 1 inch | 1 inch | 1 inch | 1 inch | 1 inch | 1 inch | 1 inch | 1 inch |

STEP 3 Compare the size of 1 foot to the size of 1 inch.

1 foot

| 1 inch | 1 inch | 1 inch | 1 inch | 1 inch | 1 inch | 1 inch | 1 inch | 1 inch | 1 inch | 1 inch | 1 inch |

| 1 inch |

1 inch

Think: You need 12 inches to make 1 foot.

So, 1 foot is _____ times as long as 1 inch.

Math Talk MATHEMATICAL PRACTICES ②

Use Reasoning Explain how you know the number of inches you need to make a yard.

⚷ Example Compare measures.

Emma has 4 feet of thread. She needs 50 inches of thread to make some bracelets. How can she determine if she has enough thread to make the bracelets?

Since 1 foot is 12 times as long as 1 inch, you can write feet as inches by multiplying the number of feet by 12.

STEP 1 Make a table that relates feet and inches.

Feet	Inches
1	12
2	
3	
4	
5	

Think:

1 foot × 12 = 12 inches

2 feet × 12 = _____

3 feet × _____ = _____

4 feet × _____ = _____

5 feet × _____ = _____

STEP 2 Compare 4 feet and 50 inches.

4 feet   50 inches

Think: Write each measure in inches and compare using <, >, or =.

_____   ◯   _____

Emma has 4 feet of thread. She needs 50 inches of thread.

4 feet is _____ than 50 inches.

So, Emma _____ enough thread to make the bracelets.

MATHEMATICAL PRACTICES ❷

Represent a Problem Explain how making a table helped you solve the problem.

- What if Emma had 5 feet of thread? Would she have enough thread to make the bracelets? Explain.

_____

_____

_____

Name _____

Share and Show **MATH BOARD**

1. Compare the size of a yard to the size of a foot.
   Use a model to help.

Customary Units of Length
1 foot (ft) = 12 inches (in.)
1 yard (yd) = 3 feet
1 yard (yd) = 36 inches

```
┌──────────────────────────┐
│                          │
│          1 yard          │
│                          │
└──────────────────────────┘
┌────────┬────────┬────────┐
│        │        │        │
│ _____ │ _____ │ _____ │
└────────┴────────┴────────┘
```

1 yard is _____ times as long as _____ foot.

Complete.

✓ 2. 2 feet = _____ inches   3. 3 yards = _____ feet   ✓ 4. 7 yards = _____ feet

Math Talk

MATHEMATICAL PRACTICES ④

Interpret a Result If you measured the length of your classroom in yards and then in feet, which unit would have a greater number of units? Explain.

On Your Own

Complete.

5. 4 yards = _____ feet   6. 10 yards = _____ feet   7. 7 feet = _____ inches

MATHEMATICAL PRACTICE ④ Use Symbols **Algebra** Compare using <, >, or =.

8. 1 foot ⬡ 13 inches   9. 2 yards ⬡ 6 feet   10. 6 feet ⬡ 60 inches

Problem Solving • Applications *Real World*

11. **THINK SMARTER** Joanna has 3 yards of fabric. She needs 100 inches of fabric to make curtains. Does she have enough fabric to make curtains? Explain. Make a table to help.

_____

_____

_____

Yards	Inches
1	
2	
3	

12. **THINK SMARTER** Select the measures that are equal. Mark all that apply.

Ⓐ 4 feet   Ⓒ 36 feet   Ⓔ 15 feet

Ⓑ 12 yards   Ⓓ 480 inches   Ⓕ 432 inches

13. **GO DEEPER** Jasmine and Luke used fraction strips to compare the size of a foot to the size of an inch using fractions. They drew models to show their answers. Whose answer makes sense? Whose answer is nonsense? Explain your reasoning.

Jasmine's Work	Luke's Work

Jasmine's Work

1

| $\frac{1}{12}$ | $\frac{1}{12}$ | $\frac{1}{12}$ | $\frac{1}{12}$ | $\frac{1}{12}$ | $\frac{1}{12}$ | $\frac{1}{12}$ | $\frac{1}{12}$ | $\frac{1}{12}$ | $\frac{1}{12}$ | $\frac{1}{12}$ | $\frac{1}{12}$ |

1 inch is $\frac{1}{12}$ of a foot.

Luke's Work

1

| $\frac{1}{3}$ | $\frac{1}{3}$ | $\frac{1}{3}$ |

1 inch is $\frac{1}{3}$ of a foot.

a. **MATHEMATICAL PRACTICE ③ Apply** For the answer that is nonsense, write an answer that makes sense.

b. Look back at Luke's model. Which two units could you compare using his model? Explain.

Name _____

Customary Units of Length

Common Core Standard—4.MD.A.1
Solve problems involving measurement and conversion of measurements from a larger unit to a smaller unit.

Complete.

1. 3 feet = ___36___ inches   Think: 1 foot = 12 inches,
   so 3 feet = 3 × 12 inches, or 36 inches

2. 2 yards = _____ feet

3. 8 feet = _____ inches

4. 7 yards = _____ feet

5. 4 feet = _____ inches

6. 15 yards = _____ feet

7. 10 feet = _____ inches

Compare using <, >, or =.

8. 3 yards ◯ 10 feet

9. 5 feet ◯ 60 inches

10. 8 yards ◯ 20 feet

Problem Solving · Real World

11. Carla has two lengths of ribbon. One ribbon is 2 feet long. The other ribbon is 30 inches long. Which length of ribbon is longer? **Explain.**

    _____

    _____

    _____

12. A football player gained 2 yards on one play. On the next play, he gained 5 feet. Was his gain greater on the first play or the second play? **Explain.**

    _____

    _____

    _____

13. **WRITE** ▸*Math* Write a problem that can be solved by comparing feet and inches using a model. Include a solution. Explain why you are changing from a larger unit to a smaller unit.

    _____

    _____

Lesson Check (4.MD.A.1)

1. Marta has 14 feet of wire to use to make necklaces. She needs to know the length in inches so she can determine how many necklaces to make. How many inches of wire does Marta have?

2. Jarod bought 8 yards of ribbon. He needs 200 inches to use to make curtains. How many inches of ribbon does he have?

Spiral Review (4.NF.C.6, 4.MD.A.1, 4.MD.A.2, 4.MD.C.5a)

3. Describe the turn shown below. (Be sure to include both the size and direction of the turn in your answer.)

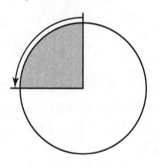

4. What decimal represents the shaded part of the model below?

5. Three sisters shared $3.60 equally. How much did each sister get?

6. Which is the best estimate for the width of your index finger?

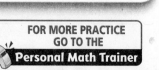

FOR MORE PRACTICE
GO TO THE
Personal Math Trainer

Name _____

Customary Units of Weight

Essential Question How can you use models to compare customary units of weight?

Common Core Measurement and Data—4.MD.A.1
Also 4.MD.A.2
MATHEMATICAL PRACTICES
MP2, MP4, MP6

Unlock the Problem (Real World)

Ounces and **pounds** are customary units of weight. How does the size of a pound compare to the size of an ounce?

Activity

Materials ■ color pencils

The number line below shows the relationship between pounds and ounces.

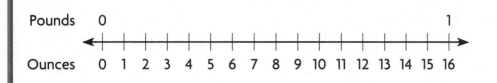

▲ You can use a spring scale to measure weight.

STEP 1 Use a color pencil to shade 1 pound on the number line.
_____

STEP 2 Use a different color pencil to shade 1 ounce on the number line.
_____

STEP 3 Compare the size of 1 pound to the size of 1 ounce.

You need _____ ounces to make _____ pound.

So, 1 pound is _____ times as heavy as 1 ounce.

Math Talk — MATHEMATICAL PRACTICES ⑥

Attend to Precision How can you compare the size of 9 pounds to the size of 9 ounces?

• **MATHEMATICAL PRACTICE ⑥ Explain** how the number line helped you to compare the sizes of the units.

_____

_____

_____

_____

🔑 Example — Compare measures.

Nancy needs 5 pounds of flour to bake pies for a festival. She has 90 ounces of flour. How can she determine if she has enough flour to bake the pies?

STEP 1 Make a table that relates pounds and ounces.

Pounds	Ounces
1	16
2	
3	
4	
5	

Think:

1 pound × 16 = 16 ounces

2 pounds × 16 = _____

3 pounds × _____ = _____

4 pounds × _____ = _____

5 pounds × _____ = _____

STEP 2 Compare 90 ounces and 5 pounds.

90 ounces   5 pounds

↓   ↓

Think: Write each measure in ounces and compare using <, >, or =.

_____ ◯ _____

Nancy has 90 ounces of flour. She needs 5 pounds of flour.

90 ounces is _____ than 5 pounds.

So, Nancy _____ enough flour to make the pies.

Try This! There are 2,000 pounds in 1 **ton**.
Make a table that relates tons and pounds.

Tons	Pounds
1	2,000
2	
3	

1 ton is _____ times as heavy as 1 pound.

Name _____

Share and Show MATH BOARD

1. 4 tons = _____ pounds

Think: 4 tons × _____ = _____

Complete.

✓ **2.** 5 tons = _____ pounds

3. 6 pounds = _____ ounces

Math Talk

MATHEMATICAL PRACTICES ④

Write an Equation What equation can you use to solve Exercise 4? Explain.

On Your Own

Complete.

✓ **4.** 7 pounds = _____ ounces

5. 6 tons = _____ pounds

MATHEMATICAL PRACTICE ④ Use Symbols **Algebra** Compare using >, <, or =.

6. 1 pound ◯ 15 ounces

7. 2 tons ◯ 2 pounds

Problem Solving • Applications Real World

8. A landscaping company ordered 8 tons of gravel. It sells the gravel in 50-pound bags. How many pounds of gravel did the company order?

_____

9. *THINK SMARTER* If you could draw a number line that shows the relationship between tons and pounds, what would it look like? Explain.

_____

_____

_____

10. *THINK SMARTER* Write the symbol that compares the weights correctly.

<	=	>

160 ounces _____ 10 pounds   600 pounds _____ 3 tons

11.  Alexis bought $\frac{1}{2}$ pound of grapes. How many ounces of grapes did she buy?

Dan drew the number line below to solve the problem. He says his model shows that there are 5 ounces in $\frac{1}{2}$ pound. What is his error?

Look at the way Dan solved the problem. Find and describe his error.

Draw a correct number line and solve the problem.

So, Alexis bought _____ ounces of grapes.

• MATHEMATICAL PRACTICE 6 Look back at the number line you drew. How many ounces are in $\frac{1}{4}$ pound? **Explain.**

Customary Units of Weight

Common Core **Common Core Standard—4.MD.A.1**
*Solve problems involving measurement and
conversion of measurements from a larger
unit to a smaller unit.*

Complete.

1. 5 pounds = _____80_____ ounces

    Think: 1 pound = 16 ounces, so
    5 pounds = 5 × 16 ounces, or 80 ounces

2. 7 tons = _____ pounds

3. 2 pounds = _____ ounces

4. 3 tons = _____ pounds

5. 10 pounds = _____ ounces

Compare using <, >, or =.

6. 8 pounds ◯ 80 ounces

7. 1 ton ◯ 100 pounds

8. 3 pounds ◯ 50 ounces

9. 5 tons ◯ 1,000 pounds

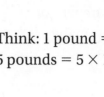 **Problem Solving** *Real World*

10. A company that makes steel girders can
produce 6 tons of girders in one day.
How many pounds is this?

11. Larry's baby sister weighed 6 pounds
at birth. How many ounces did the
baby weigh?

_____

_____

12. **WRITE** ▸*Math* Write a problem that can be solved by
comparing pounds and ounces using a model. Include a
solution. Explain why you are changing from a larger unit to a
smaller unit.

_____

_____

_____

Lesson Check (4.MD.A.1)

1. Ann bought 2 pounds of cheese to make lasagna. The recipe gives the amount of cheese needed in ounces. How many ounces of cheese did she buy?

2. A school bus weighs 7 tons. The weight limit for a bridge is given in pounds. What is this weight of the bus in pounds?

Spiral Review (4.NF.B.4c, 4.MD.A.1, 4.MD.C.7, 4.G.A.3)

3. What is the measure of $\angle EHG$?

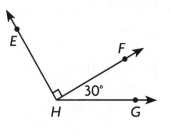

4. How many lines of symmetry does the square below have?

5. To make dough, Reba needs $2\frac{1}{2}$ cups of flour. How much flour does she need to make 5 batches of dough?

6. Judi's father is 6 feet tall. The minimum height to ride a rollercoaster is given in inches. How many inches tall is Judi's father?

FOR MORE PRACTICE
GO TO THE
Personal Math Trainer

Name _____

Customary Units of Liquid Volume

Essential Question How can you use models to compare customary units of liquid volume?

Common Core Measurement and Data—4.MD.A.1
Also 4.MD.A.2
MATHEMATICAL PRACTICES
MP3, MP4, MP6, MP7

Unlock the Problem Real World

Liquid volume is the measure of the space a liquid occupies. Some basic units for measuring liquid volume are **gallons**, **half gallons**, **quarts**, **pints**, and **cups**.

The bars below model the relationships among some units of liquid volume. The largest units are gallons. The smallest units are **fluid ounces**.

1 cup 🥛 = 8 fluid ounces
1 pint = 2 cups 🥛🥛
1 quart = 4 cups 🥛🥛🥛🥛

1 gallon

1 gallon															
1 half gallon								1 half gallon							
1 quart				1 quart				1 quart				1 quart			
1 pint		1 pint		1 pint		1 pint		1 pint		1 pint		1 pint		1 pint	
1 cup	1 cup	1 cup	1 cup	1 cup	1 cup	1 cup	1 cup	1 cup	1 cup	1 cup	1 cup	1 cup	1 cup	1 cup	1 cup
8 fluid ounces	8 fluid ounces	8 fluid ounces	8 fluid ounces	8 fluid ounces	8 fluid ounces	8 fluid ounces	8 fluid ounces	8 fluid ounces	8 fluid ounces	8 fluid ounces	8 fluid ounces	8 fluid ounces	8 fluid ounces	8 fluid ounces	8 fluid ounces

Example How does the size of a gallon compare to the size of a quart?

Math Talk MATHEMATICAL PRACTICES ⑦

Look for a Pattern
Describe the pattern in the units of liquid volume.

STEP 1 Draw two bars that represent this relationship. One bar should show gallons and the other bar should show quarts.

STEP 2 Shade 1 gallon on one bar and shade 1 quart on the other bar.

STEP 3 Compare the size of 1 gallon to the size of 1 quart.

So, 1 gallon is _____ times as much as 1 quart.

▮ **Example** Compare measures.

Serena needs to make 3 gallons of lemonade for the lemonade sale. She has a powder mix that makes 350 fluid ounces of lemonade. How can she decide if she has enough powder mix?

STEP 1 Use the model on page 659. Find the relationship between gallons and fluid ounces.

1 gallon = _____ cups

1 cup = _____ fluid ounces

1 gallon = _____ cups × _____ fluid ounces

1 gallon = _____ fluid ounces

STEP 2 Make a table that relates gallons and fluid ounces.

Gallons	Fluid Ounces
1	128
2	
3	

Think:

1 gallon = 128 fluid ounces

2 gallons × 128 = _____ fluid ounces

3 gallons × 128 = _____ fluid ounces

STEP 3 Compare 350 fluid ounces and 3 gallons.

350 fluid ounces   3 gallons

Think: Write each measure in fluid ounces and compare using <, >, or =.

_____ ◯ _____

Serena has enough mix to make 350 fluid ounces. She needs to make 3 gallons of lemonade.

350 fluid ounces is _____ than 3 gallons.

So, Serena _____ enough mix to make 3 gallons of lemonade.

Name _____

Share and Show MATH BOARD

1. Compare the size of a quart to the size of a pint.
   Use a model to help.

1 quart	

_____	_____

Customary Units of Liquid Volume
1 cup (c) = 8 fluid ounces (fl oz)
1 pint (pt) = 2 cups
1 quart (qt) = 2 pints
1 quart (qt) = 4 cups
1 gallon (gal) = 4 quarts
1 gallon (gal) = 8 pints
1 gallon (gal) = 16 cups

1 quart is _____ times as much as _____ pint.

Complete.

✓ 2. 2 pints = _____ cups

3. 3 gallons = _____ quarts

✓ 4. 6 quarts = _____ cups

On Your Own

Math Talk MATHEMATICAL PRACTICES ⑥

Make Connections Explain how the conversion chart above relates to the bar model in Exercise 1.

Use a model or *i*Tools to complete.

5. 4 gallons = _____ pints

6. 5 cups = _____ fluid ounces

MATHEMATICAL PRACTICE ④ Use Symbols **Algebra** Compare using >, <, or =.

7. 2 gallons ◯ 32 cups

8. 4 pints ◯ 6 cups

9. 5 quarts ◯ 11 pints

Problem Solving • Applications Real World

10. **THINK SMARTER** A soccer team has 25 players. The team's thermos holds 4 gallons of water. If the thermos is full, is there enough water for each player to have 2 cups? Explain. Make a table to help.

_____

_____

_____

Gallons	Cups
1	
2	
3	
4	

11. (MATHEMATICAL PRACTICE ③) **Verify the Reasoning of Others** Whose statement makes sense? Whose statement is nonsense? Explain your reasoning.

1 pint is $\frac{1}{4}$ of a gallon.

1 pint is $\frac{1}{8}$ of a gallon.

Zach's Statement	Angela's Statement

12. (GO DEEPER) Peter's glasses each hold 8 fluid ounces. How many glasses of juice can Peter pour from a bottle that holds 2 quarts?

_____

13. (THINK SMARTER) A pitcher contains 5 quarts of water. Josy says the pitcher contains 10 cups of water. Explain Josy's error. Then find the correct number of cups the pitcher contains.

_____

_____

_____

Customary Units of Liquid Volume

Common Core **Common Core Standard—4.MD.A.1**
Solve problems involving measurement and conversion of measurements from a larger unit to a smaller unit.

Complete.

1. 6 gallons = __24__ quarts

   Think: 1 gallon = 4 quarts,
   so 6 gallons = 6 × 4 quarts, or 24 quarts

2. 12 quarts = _____ pints

3. 6 cups = _____ fluid ounces

4. 9 pints = _____ cups

5. 10 quarts = _____ cups

6. 5 gallons = _____ pints

7. 3 gallons = _____ cups

Compare using <, >, or =.

8. 6 pints ◯ 60 fluid ounces

9. 3 gallons ◯ 30 quarts

10. 5 quarts ◯ 20 cups

11. 12 pints ◯ 6 cups

Problem Solving Real World

12. A chef makes $1\frac{1}{2}$ gallons of soup in a large pot. How many 1-cup servings can the chef get from this large pot of soup?

13. Kendra's water bottle contains 2 quarts of water. She wants to add drink mix to it, but the directions for the drink mix give the amount of water in fluid ounces. How many fluid ounces are in her bottle?

_____

_____

14. **WRITE** ▸*Math* Write a problem that can be solved by comparing quarts and cups using a model. Include a solution. Explain why you are changing from a larger unit to a smaller unit.

_____

_____

Lesson Check (4.MD.A.1)

1. Joshua drinks 8 cups of water a day. The recommended daily amount is given in fluid ounces. How many fluid ounces of water does he drink each day?

_____

2. A cafeteria used 5 gallons of milk in preparing lunch. How many 1-quart containers of milk did the cafeteria use?

_____

Spiral Review (4.NF.B.4a, 4.NF.C.6, 4.MD.A.1, 4.G.A.1)

3. Roy uses $\frac{1}{4}$ cup of batter for each muffin. Make a list to show the amounts of batter he will use depending on the number of muffins he makes.

_____

_____

4. Beth has $\frac{7}{100}$ of a dollar. What is the amount of money Beth has?

_____

_____

5. Name the figure that Enrico drew below.

6. A hippopotamus weighs 4 tons. Feeding instructions are given for weights in pounds. How many pounds does the hippopotamus weigh?

_____

_____

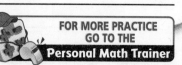

FOR MORE PRACTICE
GO TO THE
Personal Math Trainer

Name _____

Line Plots

Essential Question How can you make and interpret line plots with fractional data?

Common Core **Measurement and Data—4.MD.B.4**
Also 4.MD.A.2
MATHEMATICAL PRACTICES
MP2, MP3, MP4

Unlock the Problem (Real World)

The data show the lengths of the buttons in Jen's collection. For an art project, she wants to know how many buttons are longer than $\frac{1}{4}$ inch.

You can use a line plot to solve the problem. A **line plot** is a graph that shows the frequency of data along a number line.

Length of Buttons in Jen's Collection (in inches)
$\frac{1}{4}, \frac{3}{4}, \frac{1}{4}, \frac{4}{4}, \frac{1}{4}, \frac{4}{4}$

Make a line plot to show the data.

Example 1

STEP 1 Order the data from least to greatest length and complete the tally table.

_____

_____

STEP 2 Label the fraction lengths on the number line below from the least value of the data to the greatest.

STEP 3 Plot an *X* above the number line for each data point. Write a title for the line plot.

Buttons in Jen's Collection	
Length (in inches)	Tally
$\frac{1}{4}$	
$\frac{3}{4}$	
$\frac{4}{4}$	

_____

So, _____ buttons are longer than $\frac{1}{4}$ inch.

Math Talk MATHEMATICAL PRACTICES ④

Use Models Explain how you labeled the numbers on the number line in Step 2.

Think: To find the difference, subtract the numerators. The denominators stay the same.

1. How many buttons are in Jen's collection? _____

2. What is the difference in length between the longest button and the shortest button in Jen's collection? _____

🔓 Example 2

Some of the students in Ms. Lee's class walk to school. The data show the distances these students walk. What distance do most students walk?

Distance Students Walk to School (in miles)
$\frac{1}{2}, \frac{1}{2}, \frac{1}{4}, \frac{3}{4}, \frac{1}{4}, \frac{1}{2}, \frac{1}{2}$

Make a line plot to show the data.

STEP 1 Order the data from least to greatest distance and complete the tally table.

_____

STEP 2 Label the fraction lengths on the number line below from the least value of the data to the greatest.

STEP 3 Plot an *X* above the number line for each data point. Write a title for the line plot.

Distance Students Walk to School	
Distance (in miles)	Tally

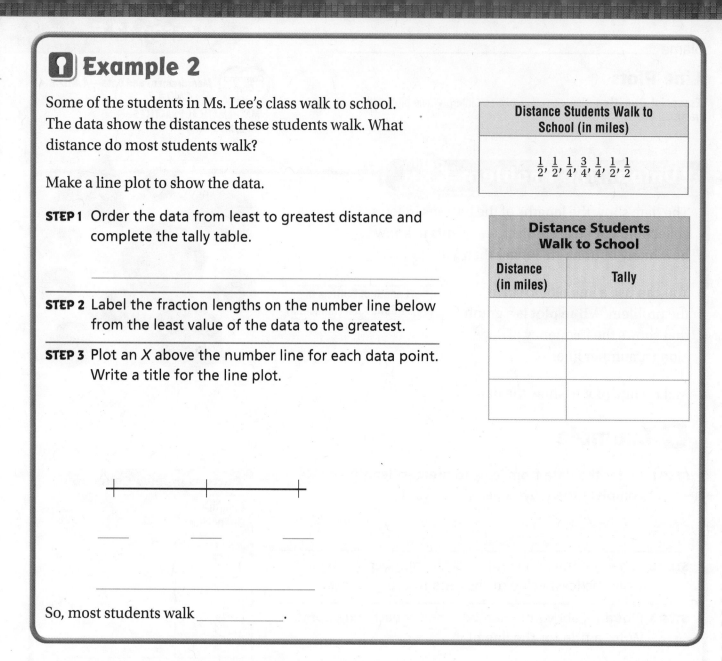

So, most students walk _____.

---

3. How many more students walk $\frac{1}{2}$ mile than $\frac{1}{4}$ mile to school?

_____

4. What is the difference between the longest distance and the shortest distance that students walk?

_____

5. What if a new student joins Ms. Lee's class who walks $\frac{3}{4}$ mile to school? How would the line plot change? Explain.

_____

Name _____

Share and Show

1. A food critic collected data on the lengths of time customers waited for their food. Order the data from least to greatest time. Make a tally table and a line plot to show the data.

Time Customers Waited for Food (in hours)
$\frac{1}{2}$, $\frac{1}{4}$, $\frac{3}{4}$, $\frac{1}{4}$, $\frac{1}{4}$, $\frac{1}{2}$, 1

Time Customers Waited for Food	
Time (in hours)	Tally

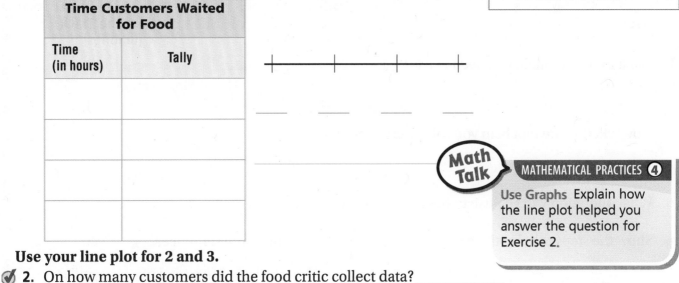

> **Math Talk**
>
> MATHEMATICAL PRACTICES ④
>
> **Use Graphs** Explain how the line plot helped you answer the question for Exercise 2.

Use your line plot for 2 and 3.

2. On how many customers did the food critic collect data? _____

3. What is the difference between the longest time and the shortest time that customers waited? _____

On Your Own

4. **MATHEMATICAL PRACTICE ④** **Use Models** The data show the lengths of the ribbons Mia used to wrap packages. Make a tally table and a line plot to show the data.

Ribbon Used to Wrap Packages	
Length (in yards)	Tally

Ribbon Length Used to Wrap Packages (in yards)
$\frac{1}{6}$, $\frac{2}{6}$, $\frac{5}{6}$, $\frac{3}{6}$, $\frac{2}{6}$, $\frac{6}{6}$, $\frac{3}{6}$, $\frac{2}{6}$

5. What is the difference in length between the longest ribbon and the shortest ribbon Mia used? _____

🔑 Unlock the Problem (Real World)

6. **GO DEEPER** The line plot shows the distances the students in Mr. Boren's class ran at the track in miles. Altogether, did the students run more or less than 5 miles?

$$\begin{array}{ccccc} X & X & & X & \\ X & X & X & X & X \end{array}$$

$$\frac{1}{5} \quad \frac{2}{5} \quad \frac{3}{5} \quad \frac{4}{5} \quad \frac{5}{5}$$

Distance Students Ran at the Track (in miles)

a. What are you asked to find? _____

_____

b. What information do you need to use? _____

_____

c. How will the line plot help you solve the problem? _____

_____

d. What operation will you use to solve the problem? _____

e. Show the steps to solve the problem.

_____

_____

f. Complete the sentences.

The students ran a total of _____ miles.

The distance is _____ than 5 miles. Altogether the students ran _____ than 5 miles.

7. **THINK SMARTER** Lena collects antique spoons. The line plot shows the lengths of the spoons in her collection. If she lines up all of her spoons in order of size, what is the size of the middle spoon? Explain.

_____

_____

$$\begin{array}{ccccc} & & & X & \\ & & & X & \\ X & & & X & \\ X & X & X & X & X \end{array}$$

$$\frac{1}{4} \quad \frac{2}{4} \quad \frac{3}{4} \quad \frac{4}{4} \quad \frac{5}{4}$$

Length of Spoons (in feet)

Personal Math Trainer

8. **THINK SMARTER +** A hiking group recorded the distances they hiked. Complete the line plot to show the data.

Distance Hiked (in miles)
$\frac{4}{8}, \frac{5}{8}, \frac{7}{8}, \frac{7}{8}, \frac{5}{8}, \frac{6}{8}, \frac{7}{8}, \frac{7}{8}, \frac{6}{8}$

Distance Hiked

Name _____

Line Plots

Common Core

COMMON CORE STANDARD—4.MD.B.4
Represent and interpret data.

1. Some students compared the time they spend riding the school bus. Complete the tally table and line plot to show the data.

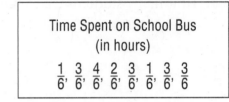

Time Spent on School Bus
(in hours)

$$\frac{1}{6}, \frac{3}{6}, \frac{4}{6}, \frac{2}{6}, \frac{3}{6}, \frac{1}{6}, \frac{3}{6}, \frac{3}{6}$$

Time Spent on School Bus	
Time (in hours)	Tally
$\frac{1}{6}$	| |
$\frac{2}{6}$	
$\frac{3}{6}$	
$\frac{4}{6}$	

**Time Spent on
School Bus (in hours)**

Use your line plot for 2 and 3.

2. How many students compared times? _____

3. What is the difference between the longest time and shortest

   time students spent riding the bus? _____

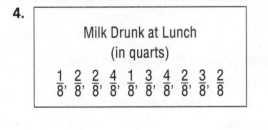

**For 4, make a tally table on a separate sheet of paper.
Make a line plot in the space below the problem.**

4.
Milk Drunk at Lunch
(in quarts)

$$\frac{1}{8}, \frac{2}{8}, \frac{2}{8}, \frac{4}{8}, \frac{1}{8}, \frac{3}{8}, \frac{4}{8}, \frac{2}{8}, \frac{3}{8}, \frac{2}{8}$$

**Milk Drunk at Lunch
(in quarts)**

5. **WRITE** ▸*Math* Write a problem that can be solved using a line plot. Draw and label the line plot and solve the problem.

_____

_____

_____

Lesson Check (4.MD.B.4)

Use the line plot for 1 and 2.

1. How many students were reading during study time?

   _____

2. What is the difference between the longest time and shortest time spent reading?

   _____

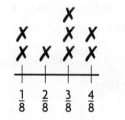

Time Spent Reading During Study Time (in hours)

Spiral Review (4.NF.C.5, 4.NF.C.6, 4.MD.A.1)

3. Bridget is allowed to play on-line games for $\frac{75}{100}$ of an hour each day. Write this fraction as a decimal.

4. Bobby's collection of sports cards has $\frac{3}{10}$ baseball cards and $\frac{39}{100}$ football cards. The rest are soccer cards. What fraction of Bobby's sports cards are baseball or football cards?

5. Jeremy gives his horse 12 gallons of water each day. How many 1-quart pails of water is that?

6. An iguana at a pet store is 5 feet long. Measurements for iguana cages are given in inches. How many inches long is the iguana?

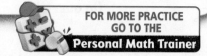

FOR MORE PRACTICE
GO TO THE
Personal Math Trainer

Name _____

✓ Mid-Chapter Checkpoint

Vocabulary

Choose the best term from the box to complete the sentence.

Vocabulary
pint
pound
yard

1. A _____ is a customary unit used to measure weight. (p. 653)

2. The cup and the _____ are both customary units for measuring liquid volume. (p. 659)

Concepts and Skills

Complete the sentence. Write *more* or *less*. (4.MD.A.1)

3. A cat weighs _____ than one ounce.

4. Serena's shoe is _____ than one yard long.

Complete. (4.MD.A.1)

5. 5 feet = _____ inches

6. 4 tons = _____ pounds

7. 4 cups = _____ pints

8. Mrs. Byrne's class went raspberry picking. The data show the weights of the cartons of raspberries the students picked. Make a tally table and a line plot to show the data. (4.MD.B.4)

Weight of Cartons of Raspberries Picked (in pounds)
$\frac{3}{4}, \frac{1}{4}, \frac{2}{4}, \frac{4}{4}, \frac{1}{4}, \frac{1}{4}, \frac{2}{4}, \frac{3}{4}, \frac{3}{4}$

Cartons of Raspberries Picked	
Weight (in pounds)	Tally

Use your line plot for 9 and 10. (4.MD.B.4)

9. What is the difference in weight between the heaviest carton and lightest carton of raspberries? _____

10. How many pounds of raspberries did Mrs. Byrne's class pick in all? _____

11. A jug contains 2 gallons of water. How many quarts of water does the jug contain? (4.MD.A.1)

_____

12. Serena bought 4 pounds of dough to make pizzas. The recipe gives the amount of dough needed for a pizza in ounces. How many ounces of dough did she buy? (4.MD.A.1)

_____

13. GO DEEPER   Vicki has a 50 inch roll of ribbon. She used 3 feet of the ribbon to wrap a gift. How many inches of ribbon does she have left? (4.MD.A.1)

_____

14. The watering can that Carlos uses in his vegetable garden holds 5 of a certain unit of liquid volume. When full, what is the best estimate for how much water is in the watering can, 5 quarts, 5 yards, or 5 ounces? (4.MD.A.1)

_____

Metric Units of Length

Essential Question How can you use models to compare metric units of length?

Measurement and Data—4.MD.A.1
Also 4.MD.A.2
MATHEMATICAL PRACTICES
MP1, MP4, MP7, MP8

Investigate

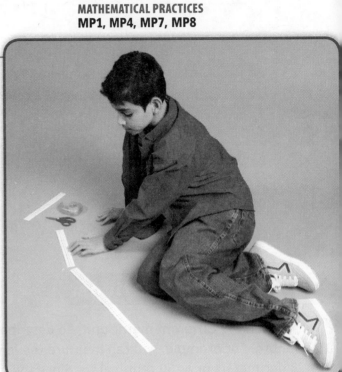

Materials ▪ ruler (meter) ▪ scissors ▪ tape

Meters (m), **decimeters** (dm), centimeters (cm), and **millimeters** (mm) are all metric units of length.

Build a meterstick to show how these units are related.

A. Cut out the meterstick strips.

B. Place the strips end-to-end to build 1 meter. Tape the strips together.

C. Look at your meter strip. What patterns do you notice about the sizes of the units?

1 meter is _____ times as long as 1 decimeter.

1 decimeter is _____ times as long as 1 centimeter.

1 centimeter is _____ times as long as 1 millimeter.

Describe the pattern you see.

_____

_____

> **Math Idea**
> If you lined up 1,000 metersticks end-to-end, the length of the metersticks would be 1 kilometer.

Draw Conclusions

1. Compare the size of 1 meter to the size of 1 centimeter. Use your meterstick to help.

_____

2. Compare the size of 1 meter to the size of 1 millimeter. Use your meterstick to help.

_____

_____

3. [THINK SMARTER] What operation could you use to find how many centimeters are in 3 meters? Explain.

_____

_____

_____

_____

Make Connections

You can use different metric units to describe the same length. For example, you can measure the length of a book as 3 decimeters or as 30 centimeters. Since the metric system is based on the number 10, decimals or fractions can be used to describe metric lengths as equivalent units.

Think of 1 meter as one whole. Use your meter strip to write equivalent units as fractions and decimals.

1 meter = 10 decimeters

Each decimeter is

_____ or _____ of a meter.

1 meter = 100 centimeters

Each centimeter is

_____ or _____ of a meter.

Complete the sentence.

- A length of 51 centimeters is _____ or _____ of a meter.

- A length of 8 decimeters is _____ or _____ of a meter.

- A length of 82 centimeters is _____ or _____ of a meter.

Math Talk

MATHEMATICAL PRACTICES **7**

Look for Structure Explain how you are able to locate and write decimeters and centimeters as parts of a meter on the meterstick.

Name _____

Metric Units of Length
1 centimeter (cm) = 10 millimeters (mm)
1 decimeter (dm) = 10 centimeters
1 meter (m) = 10 decimeters
1 meter (m) = 100 centimeters
1 meter (m) = 1,000 millimeters

Complete.

✓ **1.** 2 meters = _____ centimeters

2. 3 centimeters = _____ millimeters   **3.** 5 decimeters = _____ centimeters

MATHEMATICAL PRACTICE ④ Use Symbols Algebra Compare using <, >, or =.

4. 4 meters ◯ 40 decimeters   **5.** 5 centimeters ◯ 5 millimeters

6. 6 decimeters ◯ 65 centimeters   **7.** 7 meters ◯ 700 millimeters

**Describe the length in meters. Write your answer
as a fraction and as a decimal.**

✓ **8.** 65 centimeters = _____ or _____ meter   **9.** 47 centimeters = _____ or _____ meter

10. 9 decimeters = _____ or _____ meter   **11.** 2 decimeters = _____ or _____ meter

Problem Solving • Applications   Real World

12. A new building is 25 meters tall. How many decimeters tall is
the building?

_____

13. GO DEEPER Alexis is knitting a blanket 2 meters long. Every
2 decimeters, she changes the color of the yarn to make
stripes. How many stripes will the blanket have? Explain.

_____

_____

_____

14. **THINK SMARTER** Julianne's desk is 75 centimeters long. She says her desk is 7.5 meters long. Describe her error.

_____

_____

15. **THINK SMARTER** Write the equivalent measurements in each column.

5,000 millimeters   500 centimeters   50 centimeters

$\frac{55}{100}$ meter   0.500 meter   0.55 meter

$\frac{500}{1,000}$ meter   550 millimeters   50 decimeters

5 meters	55 centimeters	500 millimeters

16. **THINK SMARTER** Aruna was writing a report on pecan trees. She made the table of information to the right.

Write a problem that can be solved by using the data.

Pecan Tree	
Average Measurements	
Length of nuts	3 cm to 5 cm
Height	21 m to 30 m
Width of trunk	18 dm
Width of leaf	10 cm to 20 cm

Pose a problem.

Solve your problem.

- **MATHEMATICAL PRACTICE ❶** **Describe** how you could change the problem by changing a unit in the problem. Then solve the problem.

_____

_____

Metric Units of Length

Common Core

COMMON CORE STANDARD—4.MD.A.1
Solve problems involving measurement and conversion of measurements from a larger unit to a smaller unit.

Complete.

1. 4 meters = ____400____ centimeters

Think: 1 meter = 100 centimeters, so 4 meters = 4 × 100 centimeters, or 400 centimeters

2. 8 centimeters = _____ millimeters

3. 5 meters = _____ decimeters

4. 9 meters = _____ millimeters

5. 7 meters = _____ centimeters

Compare using <, >, or =.

6. 8 meters ◯ 80 centimeters

7. 3 decimeters ◯ 30 centimeters

8. 4 meters ◯ 450 centimeters

9. 90 centimeters ◯ 9 millimeters

Describe the length in meters. Write your answer as a fraction and as a decimal.

10. 43 centimeters = _____ or

   _____ meter

11. 6 decimeters = _____ **or**

   _____ meter

Problem Solving (Real World)

12. A flagpole is 4 meters tall. How many centimeters tall is the flagpole?

   _____

13. Lucille runs the 50-meter dash in her track meet. How many decimeters long is the race?

   _____

14. **WRITE** ▸*Math* Find a measurement, in centimeters, of an object. Look through books, magazines, or the Internet. Then write the measurement as parts of a meter.

   _____

   _____

Lesson Check (4.MD.A.1)

1. A pencil is 15 centimeters long. How many millimeters long is that pencil?

2. John's father is 2 meters tall. How many centimeters tall is John's father?

Spiral Review (4.NF.B.4b, 4.NF.C.7, 4.MD.B.4)

3. Bruce reads for $\frac{3}{4}$ hour each night. How long will he read in 4 nights?

4. Mark jogged 0.6 mile. Caroline jogged 0.49 mile. Write an inequality to compare the distances they jogged.

Use the line plot for 5 and 6.

5. How many lawns were mowed?

6. What is the difference between the greatest amount and least amount of gasoline used to mow lawns?

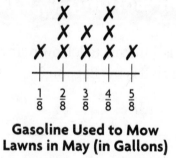

**Gasoline Used to Mow
Lawns in May (in Gallons)**

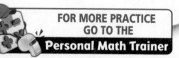

**FOR MORE PRACTICE
GO TO THE
Personal Math Trainer**

Name _____

Metric Units of Mass and Liquid Volume

Essential Question How can you compare metric units of mass and liquid volume?

Common Core **Measurement and Data—4.MD.A.1**
Also 4.MD.A.2
MATHEMATICAL PRACTICES
MP3, MP4, MP7

🔑 Unlock the Problem (Real World)

Mass is the amount of matter in an object. Metric units of mass include kilograms (kg) and grams (g). Liters (L) and **milliliters** (mL) are metric units of liquid volume.

The charts show the relationship between these units.

Metric Units of Mass
1 kilogram (kg) = 1,000 grams (g)

Metric Units of Liquid Volume
1 liter (L) = 1,000 milliliters (mL)

🔓 Example 1 Compare kilograms and grams.

Becky planted a flower garden full of bluebonnets. She used 9 kilograms of soil. How many grams of soil is that?

number of kilograms		grams in 1 kilogram		total grams
9	×	1,000	=	_____

So, Becky used _____ grams of soil to plant her bluebonnets.

- Are kilograms heavier or lighter than grams?

  _____

- Will the number of grams be greater than or less than the number of kilograms?

  _____

- What operation will you use to solve the problem?

  _____

🔓 Example 2 Compare liters and milliliters.

Becky used 5 liters of water to water her bluebonnet garden. How many milliliters of water is that?

number of liters		milliliters in 1 liter		total milliliters
5	×	1,000	=	_____

So, Becky used _____ milliliters of water.

Math Talk MATHEMATICAL PRACTICES ❼

Identify Relationships Compare the size of a kilogram to the size of a gram. Then compare the size of a liter to the size of a milliliter.

© Houghton Mifflin Harcourt Publishing Company • Image Credits: (t) ©Craig Ruaux/Alam

1. There are 3 liters of water in a pitcher. How many milliliters of water are in the pitcher?

   There are _____ milliliters in 1 liter. Since I am changing

   from a larger unit to a smaller unit, I can _____ 3 by 1,000 to find the number of milliliters in 3 liters.

   So, there are _____ milliliters of water in the pitcher.

Complete.

✓ **2.** 4 liters = _____ milliliters

✓ **3.** 6 kilograms = _____ grams

> **Math Talk**
> **MATHEMATICAL PRACTICES ⑦**
> **Look for Structure** Explain how you can find the number of grams in 8 kilograms.

On Your Own

Complete.

4. 8 kilograms = _____ grams

5. 7 liters = _____ milliliters

MATHEMATICAL PRACTICE ④ **Use Symbols Algebra** Compare using <, >, or =.

6. 1 kilogram ◯ 900 grams

7. 2 liters ◯ 2,000 milliliters

MATHEMATICAL PRACTICE ⑦ **Look for a Pattern Algebra** Complete.

8.

Liters	Milliliters
1	1,000
2	
3	
	4,000
5	
6	
	7,000
8	
9	
10	

9.

Kilograms	Grams
1	1,000
2	
	3,000
4	
5	
6	
7	
	8,000
9	
10	

Name _____

Problem Solving • Applications

10. Frank wants to fill a fish tank with 8 liters of water. How many milliliters is that?

_____

11. **GO DEEPER** Kim has 3 water bottles. She fills each bottle with 1 liter of water. How many milliliters of water does she have?

_____

12. **GO DEEPER** Jared's empty backpack has a mass of 3 kilograms. He doesn't want to carry more than 7 kilograms on a trip. How many grams of equipment can Jared pack?

_____

WRITE ▸ Math
Show Your Work

13. **GO DEEPER** A large cooler contains 20 liters of iced tea and a small cooler contains 5 liters of iced tea. How many more milliliters of iced tea does the large cooler contain than the small cooler?

_____

14. **THINK SMARTER** A 500-gram bag of granola costs $4, and a 2-kilogram bag of granola costs $15. What is the least expensive way to buy 2,000 grams of granola? Explain.

_____

_____

_____

_____

15. **MATHEMATICAL PRACTICE ❸** Verify the Reasoning of Others The world's largest apple had a mass of 1,849 grams. Sue said the mass was greater than 2 kilograms. Does Sue's statement make sense? Explain.

_____

_____

_____

_____

Unlock the Problem (Real World)

16. **THINK SMARTER** Lori bought 600 grams of cayenne pepper and 2 kilograms of black pepper. How many grams of pepper did she buy in all?

Math on the Spot

black pepper   cayenne pepper

a. What are you asked to find?

_____

b. What information will you use?

_____

c. Tell how you might solve the problem.

_____

_____

_____

d. Show how you solved the problem.

_____

_____

_____

_____

e. Complete the sentences.

Lori bought _____ grams of cayenne pepper.

She bought _____ grams of black pepper.

_____ + _____ = _____ grams

So, Lori bought _____ grams of pepper in all.

---

17. **WRITE** ▸*Math* Jill has two rocks. One has a mass of 20 grams and the other has a mass of 20 kilograms. Which rock has the greater mass? Explain.

_____

_____

_____

18. **THINK SMARTER** For numbers 18a–18c, choose Yes or No to tell whether the measurements are equivalent.

18a. 5,000 grams and 5 kilograms   ○ Yes   ○ No

18b. 300 milliliters and 3 liters   ○ Yes   ○ No

18c. 8 grams and 8,000 kilograms   ○ Yes   ○ No

Metric Units of Mass and Liquid Volume

COMMON CORE STANDARDS—4.MD.A.1
4.MD.A.2 *Solve problems involving measurement and conversion of measurements from a larger unit to a smaller unit.*

Complete.

1. 5 liters = ___**5,000**___ milliliters

Think: 1 liter = 1,000 milliliters,
so 5 liters = 5 × 1,000 milliliters, or 5,000 milliliters

2. 3 kilograms = _____ grams

3. 8 liters = _____ milliliters

4. 7 kilograms = _____ grams

5. 9 liters = _____ milliliters

Compare using <, >, or =.

6. 8 kilograms ◯ 850 grams

7. 3 liters ◯ 3,500 milliliters

Problem Solving · Real World

8. Kenny buys four 1-liter bottles of water. How many milliliters of water does Kenny buy?

_____

9. Mrs. Jones bought three 2-kilogram packages of flour. How many grams of flour did she buy?

_____

10. Colleen bought 8 kilograms of apples and 2.5 kilograms of pears. How many more grams of apples than pears did she buy?

_____

11. Dave uses 500 milliliters of juice for a punch recipe. He mixes it with 2 liters of ginger ale. How many milliliters of punch does he make?

_____

12. **WRITE** ▸*Math* Write a problem that involves changing kilograms to grams. Explain how to find the solution.

_____

_____

_____

Lesson Check (4.MD.A.1, 4.MD.A.2)

1. During his hike, Milt drank 1 liter of water and 1 liter of sports drink. How many milliliters of liquid did he drink?

_____

2. Larinda cooked a 4-kilogram roast. The roast left over after the meal weighed 3 kilograms. How many grams of roast were eaten during that meal?

_____

Spiral Review (4.MD.A.1, 4.MD.C.6, 4.G.A.1)

3. Use a protractor to find the angle measure.

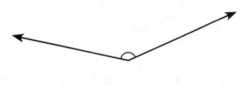

_____

4. Draw a pair of parallel lines.

5. Carly bought 3 pounds of birdseed. How many ounces of birdseed did she buy?

_____

6. A door is 8 decimeters wide. How wide is the door in centimeters?

_____

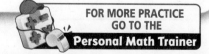
FOR MORE PRACTICE
GO TO THE
Personal Math Trainer

Name _____

Units of Time

Essential Question How can you use models to compare units of time?

Common Core   Measurement and Data—4.MD.A.1
Also 4.MD.A.2
MATHEMATICAL PRACTICES
MP1, MP4, MP5, MP7

🔑 Unlock the Problem (Real World)

The analog clock below has an hour hand, a minute hand, and a **second** hand to measure time. The time is 4:30:12.

> **Read Math**
>
> Read 4:30:12 as 4:30 and 12 seconds, or 30 minutes and 12 seconds after 4.

- Are there more minutes or seconds in one hour?

_____

There are 60 seconds in a minute and 60 minutes in an hour. The clocks show how far the hands move for each length of time.

Start Time: 3:00:00

1 second elapses.

The time is now 3:00:01.

1 minute, or 60 seconds, elapses. The second hand has made a full turn clockwise.

The time is now 3:01:00.

1 hour, or 60 minutes, elapses. The minute hand has made a full turn clockwise.

The time is now 4:00:00.

🔑 Example 1 How does the size of an hour compare to the size of a second?

There are _____ minutes in an hour.

There are _____ seconds in a minute.

60 minutes × _____ = _____ seconds

There are _____ seconds in a hour.

So, 1 hour is _____ times as long as 1 second.

Think: Multiply the number of minutes in an hour by the number of seconds in a minute.

 Math Talk

MATHEMATICAL PRACTICES ①

Analyze How many full turns clockwise does a minute hand make in 3 hours? Explain.

Chapter 12   685

🔑 Example 2 Compare measures.

Larissa spent 2 hours on her science project.
Cliff spent 200 minutes on his science project.
Who spent more time?

STEP 1 Make a table that relates hours and minutes.

Hours	Minutes
1	60
2	
3	

STEP 2 Compare 2 hours and 200 minutes.

2 hours   200 minutes

Think: Write each measure in minutes and compare using <, >, or =.

_____ ◯ _____

2 hours is _____ than 200 minutes.

So, _____ spent more time than _____ on the science project.

🔑 Activity Compare the length of a week to the length of a day.

Materials ■ color pencils

The number line below shows the relationship between days and weeks.

STEP 1 Use a color pencil to shade 1 week on the number line.

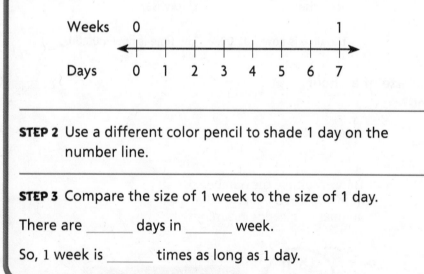

_____

STEP 2 Use a different color pencil to shade 1 day on the number line.

_____

STEP 3 Compare the size of 1 week to the size of 1 day.

There are _____ days in _____ week.

So, 1 week is _____ times as long as 1 day.

Name _____

Units of Time
1 minute (min) = 60 seconds (s)
1 hour (hr) = 60 minutes
1 day (d) = 24 hours
1 week (wk) = 7 days
1 year (yr) = 12 months (mo)
1 year (yr) = 52 weeks

1. Compare the length of a year to the length of a month. Use a model to help.

Years   0   1

Months   0 1 2 3 4 5 6 7 8 9 10 11 12

Math Talk MATHEMATICAL PRACTICES ④

Use Models Explain how the number line helped you compare the length of a year and the length of a month.

1 year is _____ times as long as _____ month.

Complete.

✓ 2. 2 minutes = _____ seconds

✓ 3. 4 years = _____ months

On Your Own

Complete.

4. 3 minutes = _____ seconds

5. 4 hours = _____ minutes

MATHEMATICAL PRACTICE ④ Use Symbols **Algebra** Compare using >, <, or =.

6. 3 years ◯ 35 months

7. 2 days ◯ 40 hours

Problem Solving • Applications Real World

8. GO DEEPER Damien has lived in the apartment building for 5 years. Ken has lived there for 250 weeks. Who has lived in the building longer? Explain. Make a table to help.

_____

_____

9. THINK SMARTER How many hours are in a week? Explain.

_____

_____

Years	Weeks
1	
2	
3	
4	
5	

Math on the Spot

10. **MATHEMATICAL PRACTICE ⑤ Communicate** Explain how you know that 9 minutes is less than 600 seconds.

_____

_____

11. **THINK SMARTER** Draw lines to match equivalent time intervals. Some intervals might not have a match.

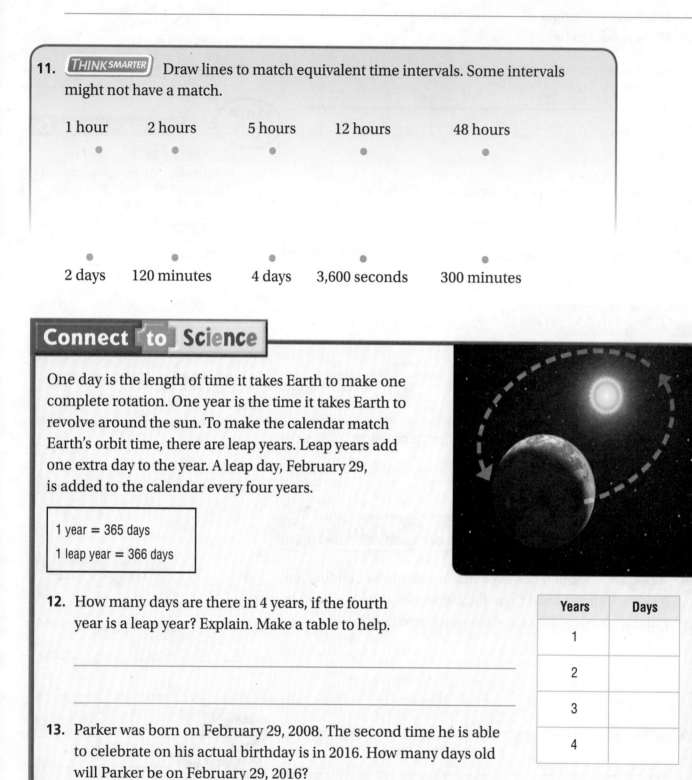

1 hour   2 hours   5 hours   12 hours   48 hours
   •   •   •   •   •

   •   •   •   •   •
2 days   120 minutes   4 days   3,600 seconds   300 minutes

Connect to Science

One day is the length of time it takes Earth to make one complete rotation. One year is the time it takes Earth to revolve around the sun. To make the calendar match Earth's orbit time, there are leap years. Leap years add one extra day to the year. A leap day, February 29, is added to the calendar every four years.

| 1 year = 365 days |
| 1 leap year = 366 days |

12. How many days are there in 4 years, if the fourth year is a leap year? Explain. Make a table to help.

_____

_____

13. Parker was born on February 29, 2008. The second time he is able to celebrate on his actual birthday is in 2016. How many days old will Parker be on February 29, 2016?

Years	Days
1	
2	
3	
4	

_____

Units of Time

 COMMON CORE STANDARD—4.MD.A.1
Solve problems involving measurement and conversion of measurements from a larger unit to a smaller unit.

Complete.

1. 6 minutes = ___360___ seconds   Think: 1 minute = 60 seconds,
                                        so 6 minutes = 6 × 60 seconds, or 360 seconds

2. 5 weeks = _____ days   3. 3 years = _____ weeks

4. 9 hours = _____ minutes   5. 9 minutes = _____ seconds

Compare using <, >, or =.

6. 2 years ◯ 14 months   7. 3 hours ◯ 300 minutes

8. 2 days ◯ 48 hours   9. 6 years ◯ 300 weeks

Problem Solving (Real World)

10. Jody practiced a piano piece for 500 seconds. Bill practiced a piano piece for 8 minutes. Who practiced longer? **Explain.**

    _____

    _____

    _____

11. Yvette's younger brother just turned 3 years old. Fred's brother is now 30 months old. Whose brother is older? **Explain.**

    _____

    _____

    _____

12. **WRITE** ▶*Math* Explain how you can prove that 3 weeks is less than 24 days.

    _____

    _____

    _____

Lesson Check (4.MD.A.1)

1. Glen rode his bike for 2 hours. For how many minutes did Glen ride his bike?

2. Tina says that vacation starts in exactly 4 weeks. In how many days does vacation start?

Spiral Review (4.NF.B.3b, 4.NF.C.5, 4.MD.A.1, 4.MD.A.2)

3. Kayla bought $\frac{9}{4}$ pounds of apples. What is that weight as a mixed number?

4. Judy, Jeff, and Jim each earned $5.40 raking leaves. How much did they earn together?

5. Melinda rode her bike $\frac{54}{100}$ mile to the library. Then she rode $\frac{4}{10}$ mile to the store. How far did Melinda ride her bike in all? Write your answer as a decimal.

6. One day, the students drank 60 quarts of milk at lunch. How many pints of milk did the students drink?

FOR MORE PRACTICE
GO TO THE
Personal Math Trainer

Name _____

Problem Solving • Elapsed Time

Essential Question How can you use the strategy *draw a diagram* to solve elapsed time problems?

Common Core   **Measurement and Data—4.MD.A.2**
Also 4.MD.A.1
MATHEMATICAL PRACTICES
MP1, MP4, MP5

Unlock the Problem   Real World

Dora and her brother Kyle spent 1 hour and 35 minutes doing yard work. Then they stopped for lunch at 1:20 P.M. At what time did they start doing yard work?

Use the graphic organizer to help you solve the problem.

Read the Problem

What do I need to find?	**What information do I need to use?**	**How will I use the information?**
I need to find the time that Dora and Kyle _____.	I need to use the _____ and the time that they _____.	I can draw a time line to help me count backward and find the _____.

Solve the Problem

I draw a time line that shows the end time 1:20 P.M. Next, I count backward 1 hour and then 5 minutes at a time until I have 35 minutes.

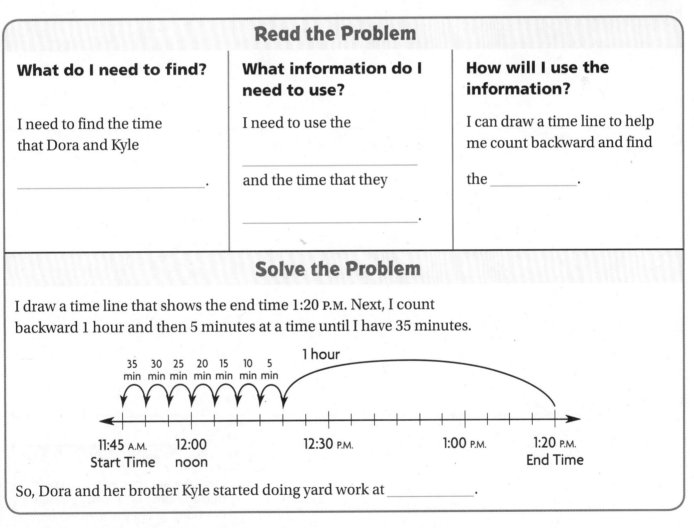

So, Dora and her brother Kyle started doing yard work at _____.

1. What if Dora and Kyle spent 50 minutes doing yard work and they stopped for lunch at 12:30 P.M.? What time would they have started doing yard work?

_____

🔓 Try Another Problem

Ben started riding his bike at 10:05 A.M. He stopped 23 minutes later when his friend Robbie asked him to play kickball. At what time did Ben stop riding his bike?

Read the Problem

What do I need to find?	What information do I need to use?	How will I use the information?

Solve the Problem

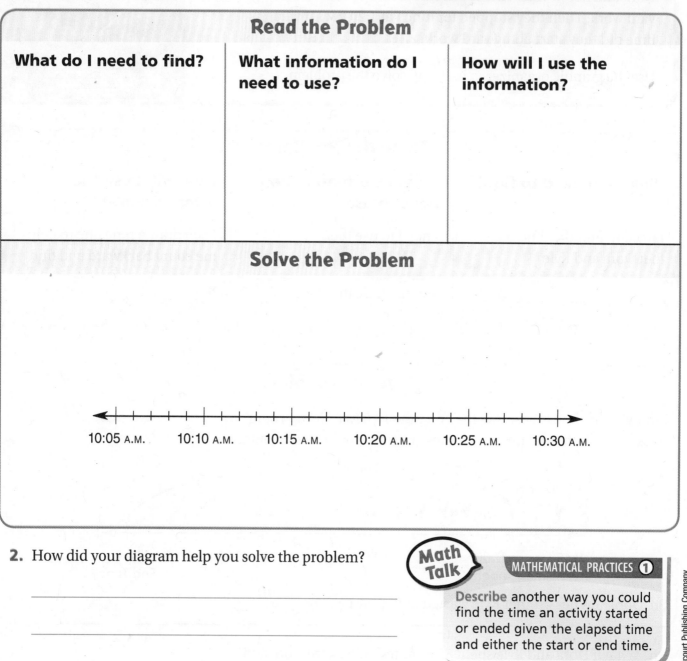

10:05 A.M.   10:10 A.M.   10:15 A.M.   10:20 A.M.   10:25 A.M.   10:30 A.M.

2. How did your diagram help you solve the problem?

_____

_____

_____

© Houghton Mifflin Harcourt Publishing Company

Math Talk

MATHEMATICAL PRACTICES ①

Describe another way you could find the time an activity started or ended given the elapsed time and either the start or end time.

Name _____

Unlock the Problem

√ Use the Problem Solving MathBoard.
√ Choose a strategy you know.
√ Underline important facts.

Share and Show MATH BOARD

1. Evelyn has dance class every Saturday. It lasts
   1 hour and 15 minutes and is over at 12:45 P.M.
   At what time does Evelyn's dance class begin?

   First, write the problem you need to solve.

   _____

   Next, draw a time line to show the end time and
   the elapsed time.

11:00 A.M.   12:00   1:00 P.M.
                             noon

   Finally, find the start time.

   Evelyn's dance class begins at _____ .

2. **THINK SMARTER** What if Evelyn's dance class started
   at 11:00 A.M. and lasted 1 hour and 25 minutes?
   At what time would her class end? Describe how
   this problem is different from Problem 1.

   _____

   _____

3. Beth got on the bus at 8:06 A.M.
   Thirty-five minutes later, she arrived
   at school. At what time did Beth arrive
   at school?

   _____

   _____

4. Lyle went fishing for 1 hour and
   30 minutes until he ran out of bait
   at 6:40 P.M. At what time did Lyle
   start fishing?

   _____

   _____

On Your Own

5. Mike and Jed went skiing at 10:30 A.M. They skied for 1 hour and 55 minutes before stopping for lunch. At what time did Mike and Jed stop for lunch?

_____

6. GO DEEPER Mike can run a mile in 12 minutes. He starts his run at 11:30 AM. and runs 4 miles. What time does Mike finish his run?

_____

7. MATHEMATICAL PRACTICE 5 Communicate Explain how you can use a diagram to determine the start time when the end time is 9:00 A.M. and the elapsed time is 26 minutes. What is the start time?

_____

_____

_____

WRITE ▸ Math
Show Your Work

8. THINK SMARTER Bethany finished her math homework at 4:20 P.M. She did 25 multiplication problems in all. If each problem took her 3 minutes to do, at what time did Bethany start her math homework?

_____

9. THINK SMARTER Vincent began his weekly chores on Saturday morning at 11:20 A.M. He finished 1 hour and 10 minutes later. Draw a time line to show the end time.

◄——┼—┼—┼—┼—┼—┼—┼—┼—┼—┼—┼—┼—┼—┼—┼—┼—┼—┼——►

11:00 A.M.   12:00   1:00 P.M.
                               noon

Vincent finished his chores at _____ P.M.

Problem Solving • Elapsed Time

Common Core **COMMON CORE STANDARD—4.MD.A.2**
Solve problems involving measurement and conversion of measurements from a larger unit to a smaller unit.

Read each problem and solve.

1. Molly started her piano lesson at 3:45 P.M. The lesson lasted 20 minutes. What time did the piano lesson end?

   Think: What do I need to find? How can I draw a diagram to help?

   ```
        5 min    10 min    15 min    20 min
   ◄─────╮─────╮──────╮──────╮──────►
   3:45 P.M.  3:50 P.M.  3:55 P.M.  4:00 P.M.  4:05 P.M.
   Start Time                            End Time
   ```

   _____4:05 P.M._____

2. Brendan spent 24 minutes playing a computer game. He stopped playing at 3:55 P.M and went outside to ride his bike. What time did he start playing the computer game?

   _____

3. Aimee's karate class lasts 1 hour and 15 minutes and is over at 5:00 P.M. What time does Aimee's karate class start?

   _____

4. Mr. Giarmo left for work at 7:15 A.M. Twenty-five minutes later, he arrived at his work. What time did Mr. Giarmo arrive at his work?

   _____

5. **WRITE** ►*Math* Explain why it is important to know if a time is in the A.M. or in the P.M. when figuring out how much time has elapsed.

   _____

   _____

Lesson Check (4.MD.A.2)

1. Bobbie went snowboarding with friends at
   10:10 A.M. They snowboarded for 1 hour
   and 43 minutes, and then stopped to eat
   lunch. What time did they stop for lunch?

   _____

2. The Cain family drove for 1 hour and
   15 minutes and arrived at their camping
   spot at 3:44 P.M. What time did the Cain
   family start driving?

   _____

Spiral Review (4.NF.B.4b, 4.NF.C.5, 4.MD.A.1, 4.MD.A.2)

3. A praying mantis can grow up to
   15 centimeters long. How long is
   this in millimeters?

   _____

4. Thom's minestrone soup recipe makes
   3 liters of soup. How many milliliters of
   soup is this?

   _____

5. Stewart walks $\frac{2}{3}$ mile each day. List three
   multiples of $\frac{2}{3}$.

   _____

   _____

6. Angelica colored in 0.60 of the squares
   on her grid. Write 0.60 as tenths in
   fraction form.

   _____

   _____

© Houghton Mifflin Harcourt Publishing Company

FOR MORE PRACTICE
GO TO THE
Personal Math Trainer

Name _____

Mixed Measures

Essential Question How can you solve problems involving mixed measures?

Common Core Measurement and Data—4.MD.A.2
Also 4.MD.A.1
MATHEMATICAL PRACTICES
MP2, MP3, MP8

Unlock the Problem Real World

Herman is building a picnic table for a new campground. The picnic table is 5 feet 10 inches long. How long is the picnic table in inches?

• Is the mixed measure greater than or less than 6 feet?
_____
• How many inches are in 1 foot?
_____

🔓 Change a mixed measure.

Think of 5 feet 10 inches as 5 feet + 10 inches.

Write feet as inches.

$$5 \text{ feet}$$
$$+ 10 \text{ inches}$$

Think: 5 feet × 12 = ⟶ 60 inches

☐ inches
+ ☐ inches
☐ inches

So, the picnic table is _____ inches long.

🔓 Example 1 Add mixed measures.

Herman built the picnic table in 2 days. The first day he worked for 3 hours 45 minutes. The second day he worked for 2 hours 10 minutes. How long did it take him to build the table?

STEP 1 Add the minutes.

3 hr 45 min
+ 2 hr 10 min
_____
☐ min

STEP 2 Add the hours.

3 hr 45 min
+ 2 hr 10 min
_____
☐ hr 55 min

So, it took Herman _____ to build the table.

Math Talk

MATHEMATICAL PRACTICES ⑧

Use Repeated Reasoning
How is adding mixed measures similar to adding tens and ones? How is it different? Explain.

• What if Herman worked an extra 5 minutes on the picnic table? How long would he have worked on the table then? Explain.

_____

_____

🔒 Example 2 Subtract mixed measures.

Alicia is building a fence around the picnic area. She has a pole that is 6 feet 6 inches long. She cuts off 1 foot 7 inches from one end. How long is the pole now?

STEP 1 Subtract the inches.

Think: 7 inches is greater than 6 inches. You need to regroup to subtract.

6 ft 6 in. = 5 ft 6 in. + 12 in.

= 5 ft _____ in.

$$
\begin{array}{r}
\overset{5}{\cancel{6}}\,\text{ft}\;\overset{18}{\cancel{0}}\,\text{in.} \\
-\;1\,\text{ft}\;7\,\text{in.} \\
\hline
\text{in.}
\end{array}
$$

> **! ERROR Alert**
> Be sure to check that you are regrouping correctly. There are 12 inches in 1 foot.

STEP 2 Subtract the feet.

$$
\begin{array}{r}
\overset{5}{\cancel{6}}\,\text{ft}\;\overset{18}{\cancel{0}}\,\text{in.} \\
-\;1\,\text{ft}\;7\,\text{in.} \\
\hline
\text{ft}\;11\,\text{in.}
\end{array}
$$

So, the pole is now _____ long.

Try This! Subtract.

3 pounds 5 ounces − 1 pound 2 ounces

Share and Show

1. A truck is carrying 2 tons 500 pounds of steel. How many pounds of steel is the truck carrying?

Think of 2 tons 500 pounds as 2 tons + 500 pounds.
Write tons as pounds.

2 tons   Think: 2 tons × 2,000 = ⟶   _____ pounds
+ 500 pounds   _____ pounds   + _____ pounds
                                                _____ pounds

So, the truck is carrying _____ pounds of steel.

698

Name _____

Rewrite each measure in the given unit.

2. 1 yard 2 feet

_____ feet

3. 3 pints 1 cup

_____ cups

✓ **4.** 3 weeks 1 day

_____ days

Add or subtract.

5. 2 lb 4 oz
    + 1 lb 6 oz

✓ **6.** 3 gal 2 qt
    − 1 gal 3 qt

7. 5 hr 20 min
    − 3 hr 15 min

Math Talk

MATHEMATICAL PRACTICES ❷

Reason Quantitatively
How do you know when you need to regroup to subtract? Explain.

On Your Own

Rewrite each measure in the given unit.

8. 1 hour 15 minutes

_____ minutes

9. 4 quarts 2 pints

_____ pints

10. 10 feet 10 inches

_____ inches

Add or subtract.

11. 2 tons 300 lb
    − 1 ton 300 lb

12. 10 gal 8 c
    + 8 gal 9 c

13. 7 lb 6 oz
    − 2 lb 12 oz

Problem Solving • Applications Real World

14. MATHEMATICAL PRACTICE ❸ **Apply** Ahmed fills 6 pitchers with juice. Each pitcher contains 2 quarts 1 pint. How many pints of juice does he have in all?

_____

15. **Sense or Nonsense?** Sam and Dave each solve the problem at the right. Sam says the sum is 4 feet 18 inches. Dave says the sum is 5 feet 6 inches. Whose answer makes sense? Whose answer is nonsense? Explain.

  2 ft 10 in.
+ 2 ft 8 in.

_____

16. THINK SMARTER Jackson has a rope 1 foot 8 inches long. He cuts it into 4 equal pieces. How many inches long is each piece?

_____

🔑 Unlock the Problem Real World

17. Theo is practicing for a 5-kilometer race. He runs
5 kilometers every day and records his time. His normal
time is 25 minutes 15 seconds. Yesterday it took him only
23 minutes 49 seconds. How much faster was his time
yesterday than his normal time?

a. What are you asked to find?

_____

_____

b. What information do you know?

_____

_____

c. How will you solve the problem?

_____

d. Solve the problem.

_____

e. Fill in the sentence.

Yesterday, Theo ran 5 kilometers in a time

that was _____ faster than his
normal time.

18. GO DEEPER Don has 5 pieces of pipe. Each
piece is 3 feet 6 inches long. If Don joins the
pieces end to end to make one long pipe,
how long will the new pipe be?

19. THINK SMARTER + Ana mixes
2 quarts 1 pint of apple juice and 1 quart
3 cups of cranberry juice. Will her mixture
be able to fit in a 1 gallon pitcher? Explain.

_____

_____

_____

_____

Mixed Measures

 COMMON CORE STANDARD—4.MD.A.2
Solve problems involving measurement and conversion of measurements from a larger unit to a smaller unit.

Complete.

1. 8 pounds 4 ounces = _____132_____ ounces

Think: 8 pounds = 8 × 16 ounces, or 128 ounces.

128 ounces + 4 ounces = 132 ounces

2. 5 weeks 3 days = _____ days

3. 4 minutes 45 seconds = _____ seconds

4. 4 hours 30 minutes = _____ minutes

5. 3 tons 600 pounds = _____ pounds

Add or subtract.

6.   9 gal 1 qt
  + 6 gal 1 qt
  —————

7.   12 lb 5 oz
  − 7 lb 10 oz
  —————

8.   8 hr 3 min
  + 4 hr 12 min
  —————

Problem Solving (Real World)

9. Michael's basketball team practiced for 2 hours 40 minutes yesterday and 3 hours 15 minutes today. How much longer did the team practice today than yesterday?

10. Rhonda had a piece of ribbon that was 5 feet 3 inches long. She removed a 5-inch piece to use in her art project. What is the length of the piece of ribbon now?

_____

_____

11. **WRITE** ▸*Math* Write a subtraction problem involving pounds and ounces. Solve the problem and show your work.

_____

_____

_____

Lesson Check (4.MD.A.2)

1. Marsha bought 1 pound 11 ounces of roast beef and 2 pounds 5 ounces of corned beef. How much more corned beef did she buy than roast beef?

2. Theodore says there are 2 weeks 5 days left in the year. How many days are left in the year?

Spiral Review (4.NF.C.7, 4.MD.A.1, 4.MD.A.2, 4.G.A.2)

3. On one grid, 0.5 of the squares are shaded. On another grid, 0.05 of the squares are shaded. Compare the shaded parts of the grids using <, =, or >.

4. Classify the triangle by the size of its angles.

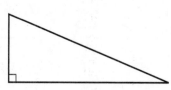

5. Sahil's brother is 3 years old. How many weeks old is his brother?

6. Sierra's swimming lessons last 1 hour 20 minutes. She finished her lesson at 10:50 A.M. At what time did her lesson start?

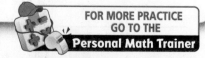

FOR MORE PRACTICE
GO TO THE
Personal Math Trainer

Name _____

Patterns in Measurement Units

Essential Question How can you use patterns to write number pairs for measurement units?

Common Core Measurement and Data—
4.MD.A.1
MATHEMATICAL PRACTICES
MP3, MP6, MP7

CONNECT The table at the right relates yards and feet. You can think of the numbers in the table as number pairs. 1 and 3, 2 and 6, 3 and 9, 4 and 12, and 5 and 15 are number pairs.

The number pairs show the relationship between yards and feet. 1 yard is equal to 3 feet, 2 yards is equal to 6 feet, 3 yards is equal to 9 feet, and so on.

Yards	Feet
1	3
2	6
3	9
4	12
5	15

Unlock the Problem *Real World*

Lillian made the table below to relate two units of time. What units of time does the pattern in the table show?

Activity Use the relationship between the number pairs to label the columns of the table.

1	7
2	14
3	21
4	28
5	35

- List the number pairs.

  _____

- Describe the relationship between the numbers in each pair.

  _____

  _____

- Label the columns of the table. **Think:** What unit of time is 7 times as great as another unit?

Math Talk MATHEMATICAL PRACTICES ⑦

Identify Relationships
Look at each number pair in the table. Could you change the order of the numbers in the number pairs? Explain why or why not.

Try This! Jasper made the table below to relate two customary units of liquid volume. What customary units of liquid volume does the pattern in the table show?

- List the number pairs.

  _____

- Describe the relationship between the numbers in each pair.

  _____

  _____

  _____

- Label the columns of the table.

_____	_____
1	4
2	8
3	12
4	16
5	20

Think: What customary unit of liquid volume is 4 times as great as another unit?

- What other units could you have used to label the columns of the table above? Explain.

  _____

  _____

  _____

Share and Show MATH BOARD

1. The table shows a pattern for two units of time. Label the columns of the table with the units of time.

   Think: What unit of time is 24 times as great as another unit?

_____	_____
1	24
2	48
3	72
4	96
5	120

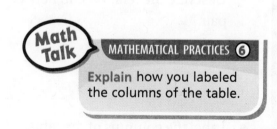

Math Talk MATHEMATICAL PRACTICES ⑥

Explain how you labeled the columns of the table.

Name _____

Each table shows a pattern for two customary units. Label the columns of the table.

✓ **2.**

____	____
1	2
2	4
3	6
4	8
5	10

✓ **3.**

____	____
1	16
2	32
3	48
4	64
5	80

On Your Own

Each table shows a pattern for two customary units. Label the columns of the table.

4.

____	____
1	36
2	72
3	108
4	144
5	180

5.

____	____
1	12
2	24
3	36
4	48
5	60

Each table shows a pattern for two metric units of length. Label the columns of the table.

6.

____	____
1	10
2	20
3	30
4	40
5	50

7.

____	____
1	100
2	200
3	300
4	400
5	500

8. **GO DEEPER** List the number pairs for the table in Exercise 6. Describe the relationship between the numbers in each pair.

_____

_____

_____

Problem Solving · Applications Real World

9. What's the Error? Maria wrote *Weeks* as the label for the first column of the table and *Years* as the label for the second column. Describe her error.

_____

_____

_____

?	?
1	52
2	104
3	156
4	208
5	260

10. **MATHEMATICAL PRACTICE ❸ Verify the Reasoning of Others** The table shows a pattern for two metric units. Lou labels the columns *Meters* and *Millimeters*. Zayna labels them *Liters* and *Milliliters*. Whose answer makes sense? Whose answer is nonsense? Explain.

_____

_____

_____

_____

?	?
1	1,000
2	2,000
3	3,000
4	4,000
5	5,000

11. _THINK SMARTER_ Look at the following number pairs: 1 and 365, 2 and 730, 3 and 1,095. The number pairs describe the relationship between which two units of time? Explain.

_____

_____

12. _THINK SMARTER_ The tables show patterns for some units of measurement. Write the correct labels in each table.

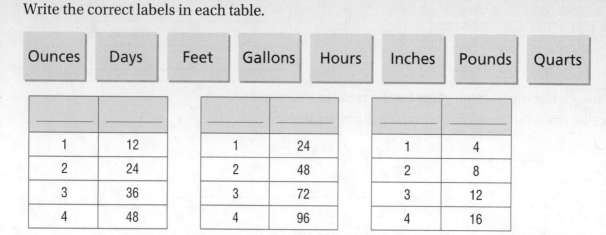

Ounces	Days	Feet	Gallons	Hours	Inches	Pounds	Quarts

___	___
1	12
2	24
3	36
4	48

___	___
1	24
2	48
3	72
4	96

___	___
1	4
2	8
3	12
4	16

Name _____

Patterns in Measurement Units

COMMON CORE STANDARD—4.MD.A.1
Solve problems involving measurement and conversion of measurements from a larger unit to a smaller unit.

Each table shows a pattern for two customary units of time, liquid volume, or weight. Label the columns of the table.

1.

Gallons	Quarts
1	4
2	8
3	12
4	16
5	20

2.

_____	_____
1	2,000
2	4,000
3	6,000
4	8,000
5	10,000

3.

_____	_____
1	2
2	4
3	6
4	8
5	10

4.

_____	_____
1	60
2	120
3	180
4	240
5	300

Problem Solving · Real World

Use the table for 5.

5. Marguerite made the table to compare two metric measures of length. Name a pair of units Marguerite could be comparing.

_____

?	?
1	10
2	20
3	30
4	40
5	50

6. **WRITE** ▸*Math* Draw a table to represent months and years. Explain how you labeled each column.

_____

_____

Lesson Check (4.MD.A.1)

1. Joanne made a table to relate two units of measure. The number pairs in her table are 1 and 16, 2 and 32, 3 and 48, 4 and 64. What are the best labels for Joanne's table?

2. Cade made a table to relate two units of time. The number pairs in his table are 1 and 24, 2 and 48, 3 and 72, 4 and 96. What are the best labels for Cade's table?

Spiral Review (4.NF.C.6, 4.MD.A.1, 4.MD.A.2, 4.MD.C.5a)

3. Anita has 2 quarters, 1 nickel, and 4 pennies. Write Anita's total amount as a fraction of a dollar.

4. The minute hand of a clock moves from 12 to 6. What describes the turn the minute hand makes?

5. Roderick has a dog that has a mass of 9 kilograms. What is the mass of the dog in grams?

6. Kari mixed 3 gallons 2 quarts of lemon-lime drink with 2 gallons 3 quarts of pink lemonade to make punch. How much more lemon-lime drink did Kari use than pink lemonade?

FOR MORE PRACTICE GO TO THE Personal Math Trainer

✓ Chapter 12 Review/Test

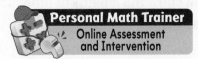

Personal Math Trainer
Online Assessment
and Intervention

1. Mrs. Miller wants to estimate the width of the steps in front of her house. Select the best benchmark for her to use.

   (A) her fingertip

   (B) the thickness of a dime

   (C) the width of a license plate

   (D) how far she can walk in 20 minutes

2. **GO DEEPER** Franco played computer chess for 3 hours. Lian played computer chess for 150 minutes. Compare the times spent playing computer chess. Complete the sentence.

   _____ played for _____ longer than _____ .

3. Select the measures that are equal. Mark all that apply.

   (A) 6 feet   (D) 600 inches

   (B) 15 yards   (E) 12 feet

   (C) 45 feet   (F) 540 inches

4. Jackie made 6 quarts of lemonade. Jackie says she made 3 pints of lemonade. Explain Jackie's error. Then find the correct number of pints of lemonade.

5. Josh practices gymnastics each day after school. The data show the lengths of time Josh practiced gymnastics for 2 weeks.

Time Practicing Gymnastics (in hours)
$\frac{1}{4}, \frac{1}{4}, \frac{3}{4}, \frac{3}{4}, \frac{1}{2}, 1, 1, 1, \frac{3}{4}, 1$

Part A

Make a tally table and line plot to show the data.

Time Practicing Gymnastics	
Time (in hours)	Tally

Part B

Explain how you used the tally table to label the numbers and plot the Xs.

Part C

What is the difference between the longest time and shortest time Josh spent practicing gymnastics?

_____ hour

6. Select the correct word to complete the sentence.

Juan brings a water bottle with him to soccer practice.

1 liter
10 milliliters
1 meter

A full water bottle holds _____ of water.

Name _____

7. Write the symbol that compares the weights correctly.

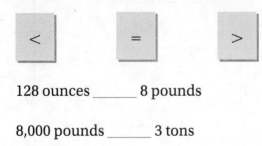

128 ounces _____ 8 pounds

8,000 pounds _____ 3 tons

8. Dwayne bought 5 yards of wrapping paper. How many inches of wrapping paper did he buy?

_____ inches

9. A sack of potatoes weighs 14 pounds 9 ounces. After Wendy makes potato salad for a picnic, the sack weighs 9 pounds 14 ounces. What is the weight of the potatoes Wendy used for the potato salad? Write the numbers to show the correct subtraction.

| 4 | 5 | 11 | 13 | 19 | 25 | 39 |

```
☐          ☐
14 pounds      9 ounces
− 9 pounds     14 ounces
☐ pounds      ☐ ounces
```

10. Sabita made this table to relate two customary units of liquid volume.

Part A

List the number pairs for the table. Then describe the relationship between the numbers in each pair.

_____	_____
1	2
2	4
3	6
4	8
5	10

Part B

Label the columns of the table. Explain your answer.

11. THINKSMARTER + The table shows the distances some students swam in miles. Complete the line plot to show the data.

Distance Students Swam(in miles)
$\frac{1}{8}, \frac{2}{8}, \frac{3}{8}, \frac{3}{8}, \frac{5}{8}, \frac{3}{8}, \frac{2}{8}, \frac{4}{8}, \frac{3}{8}, \frac{1}{8}, \frac{4}{8}, \frac{4}{8}$

Distance Students Swam (in miles)

What is the difference between the longest distance and the shortest distance the students swam?

☐ mile

12. An elephant living in a wildlife park weighs 4 tons. How many pounds does the elephant weigh?

_____ pounds

13. Katia bought two melons. She says the difference in mass between the melons is 5,000 grams. Which two melons did Katia buy?

(A) watermelon: 8 kilograms

(B) cantaloupe: 5 kilograms

(C) honeydew: 3 kilograms

(D) casaba melon: 2 kilograms

(E) crenshaw melon: 1 kilogram

14. Write the equivalent measurements in each column.

3,000 millimeters	300 centimeters	30 centimeters
$\frac{35}{100}$ meter	0.300 meter	0.35 meter
$\frac{300}{1,000}$ meter	350 millimeters	30 decimeters
3 meters	35 centimeters	300 millimeters

15. Cheryl is making a mixed fruit drink for a party. She mixes 7 pints each of apple juice and cranberry juice. How many fluid ounces of mixed fruit drink does Cheryl make?

_____ fluid ounces

16. Hamid's soccer game will start at 11:00 A.M., but the players must arrive at the field three-quarters of an hour early to warm up. The game must end by 1:15 P.M.

Part A

Hamid says he has to be at the field at 9:45 A.M. is Hamid correct? Explain your answer.

Part B

The park closes at 6:30 P.M. There is a 15-minute break between each game played at the park, and each game takes the same amount of time as Hamid's soccer game. How many more games can be played before the park closes? Explain your answer.

17. For numbers 17a–17e, select Yes or No to tell whether the measurements are equivalent.

17a. 7,000 grams and 7 kilograms   ○ Yes   ○ No

17b. 200 milliliters and 2 liters   ○ Yes   ○ No

17c. 6 grams and 6,000 kilograms   ○ Yes   ○ No

17d. 5 liters and 5,000 milliliters   ○ Yes   ○ No

17e. 2 milliliters and 2,000 liters   ○ Yes   ○ No

18. Draw lines to match equivalent time intervals.

$\frac{1}{2}$ hour   2 hours   3 hours   8 hours   72 hours

  •   •   •   •   •

  •   •   •   •   •

3 days   180 minutes   1,800 seconds   480 minutes   7,200 seconds

19. Anya arrived at the library on Saturday morning at 11:10 A.M. She left the library 1 hour 20 minutes later. Draw a time line to show the end time.

11:00 A.M.   12:00 noon   1:00 P.M.

Anya left the library at _____ P.M.

20. The tables show patterns for some units of measurement. Write the correct labels in each table.

| Pints | Days | Feet | Cups | Week | Yards | Inches | Quarts |

1	3		1	7		1	4
2	6		2	14		2	8
3	9		3	21		3	12
4	12		4	28		4	16

21. An Olympic swimming pool is 25 meters wide. How many decimeters wide is an Olympic swimming pool?

_____ decimeters wide

22. Frankie is practicing for a 5-kilometer race. His normal time is 31 minutes 21 seconds. Yesterday it took him only 29 minutes 38 seconds.

How much faster was Frankie yesterday than his normal time?

_____